Fundamentals of Engineering Metallurgy and Materials

S I metric edition

F. W. J. BAILEY

B.Sc.(Hons.), A.I.M. Senior Lecturer in the Department of Mechanical Engineering. North Gloucestershire College of Technology, Cheltenham

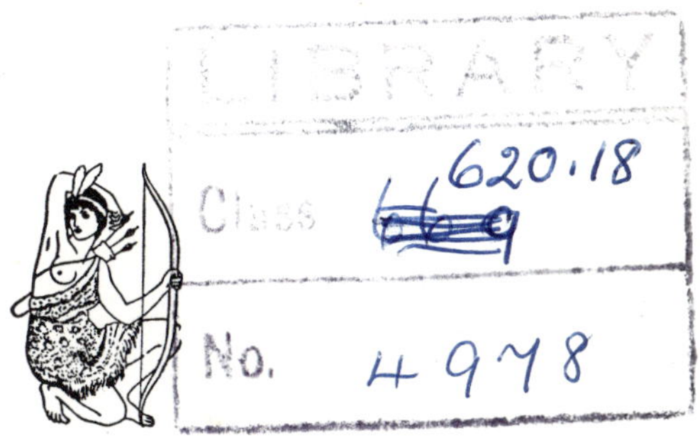

CASSELL · LONDON

CASSELL & COMPANY LTD
35 RED LION SQUARE, LONDON WC1R 4SJ
SYDNEY · AUCKLAND
TORONTO · JOHANNESBURG

First published as *Fundamentals of Engineering Metallurgy* April 1961
Second (revised and enlarged) edition July 1964
Third edition December 1965
Fourth edition September 1967
Fourth edition, second impression August 1968
Fifth (revised and enlarged) edition April 1972

I.S.B.N. 0 304 93862 9

Printed in Great Britain by
The Camelot Press Ltd., London and Southampton
472

Fundamentals of Engineering Metallurgy and Materials

CONTENTS

PREFACE TO FIFTH EDITION

In this new edition S I units have been used throughout. It is hoped that the choice of the N/mm^2 for stress will not confuse those students using the MN/m^2 as they are numerically the same. A new chapter on polymer materials has been introduced. This includes the structure and properties of polymers, composite materials, expanded plastics and foams and also a discussion of visco-elasticity. The chapters involving corrosion and fatigue have been enlarged and the opportunity taken to bring the text up to date and more in line with current syllabuses.

It is hoped that these additions will assist those students preparing for examinations in Properties of Materials in the H.N.C. and H.N.D. courses in Engineering. It should also be of assistance to students preparing for Part II of the Mechanical Engineering Technicians Course.

August 1971

PREFACE TO FIRST EDITION

The object of this book is to provide a short account of the fundamentals of Engineering Metallurgy. The author wishes to emphasize that the book has been written for the engineering student and not for the metallurgist. The study of Engineering Metallurgy requires a different approach from that of other engineering subjects, and this, together with the peculiar terminology of the metallurgist, creates special difficulties, particularly to a student with a limited background of physics and chemistry. In view of this, certain chapters have been deliberately simplified and details of metallurgical theories omitted. Wherever possible, simple diagrams have replaced long discussions.

It is hoped that students for the Higher National Certificates in Mechanical and Production Engineering will derive benefit from this treatment. Certain portions of the book should also prove useful to those studying for the Final City and Guilds of London Institute's Certificate in Machine Shop Engineering.

It is not possible to cover all aspects of the subject in a short book, and the author has, therefore, omitted details of production processes and mechanical testing.

The author wishes particularly to thank his colleague Mr. R. Kershaw, B.Sc.(Eng.) Hons., A.M.I.Mech.E., for drawing the diagrams and making many helpful suggestions.

The author wishes to thank the City and Guilds of London Institute for permission to publish past questions from the Final City and Guilds Examination in Machine Shop Engineering, and the Joint Committee for National Certificates and Diplomas in Mechanical Engineering for permission to include past questions which have been set for the Higher National Certificates in Mechanical and Production Engineering at the North Gloucestershire Technical College, Cheltenham.

Thanks are also due to Messrs. Edward Arnold (Publishers) Ltd., for the loan of a block from *Introduction to Metallic Corrosion* by Dr. U. R. Evans, and to Messrs. Kelvin and Hughes (Industrial) Ltd., for permission to reproduce diagrams from their publication *Kelvin and Hughes Ultrasonic Flaw Detector Mark 5*. Details of etching reagents for the macro- and microscopical examination of metals and alloys given in Chapter 15 have been mainly taken from *Photomicrography with the Vickers Projection Microscope* by kind permission of the publishers Messrs. Cooke, Troughton and Simms Ltd.

The author of a general textbook depends for much of his information on the work of other more specialised publications. Many of the latter are mentioned in the comprehensive bibliography and the author wishes to acknowledge their assistance, and recommends them to students for further reading. If any omission has been made, or if due acknowledgement has not been given, the author tenders his sincere apologies to the individuals or authorities concerned.

April 1961

ABBREVIATIONS USED IN THE TEXT

Ag	= Silver	Mo	= Molybdenum
Al	= Aluminium	Ni	= Nickel
As	= Arsenic	P	= Phosphorus
Be	= Beryllium	Pb	= Lead
B	= Boron	S	= Sulphur
C	= Carbon	Sb	= Antimony
Cd	= Cadmium	Se	= Selenium
Ce	= Cerium	Si	= Silicon
Co	= Cobalt	Sn	= Tin
Cr	= Chromium	Ti	= Titanium
Cu	= Copper	TC	= Total Carbon
Fe	= Iron	V	= Vanadium
Mg	= Magnesium	W	= Tungsten
Mn	= Manganese	Zn	= Zinc
		Zr	= Zirconium

TS	= Tensile Strength in newtons per square millimetre
N/mm^2	= Newtons per square millimetre
El	= Percentage Elongation on gauge length of $5{\cdot}65\sqrt{A}$
RA	= Percentage Reduction of Area
YP	= Yield Point in N/mm^2
LP	= Limit of Proportionality in N/mm^2
PS	= Proof Stress (0·1% or 0·2% as stated) in N/mm^2
HB	= Brinell Hardness Number
HV	= Vickers Hardness Number
HRC	= Rockwell Hardness Number (Scale C)
HRB	= Rockwell Hardness Number (Scale B)

IMPORTANT PHYSICAL PROPERTIES OF SOME PURE METALS

Metal	*Symbol*	*Melting Point* °C	*Relative Density* kg/m^3	*Specific Heat Capacity* $J/kg°C$	*Thermal Conductivity* $W/m°C$	*Electrical Resistivity* $n\Omega m$	*Coefficient of Thermal Expansion* $\times 10^6$ *at* 20°C
Aluminium	Al	660	2 710	875	234	28·2	24
Copper	Cu	1 083	8 960	386	384	16·7	17
Iron	Fe	1 539	7 870	438	75	100	12
Lead	Pb	327	11 340	129	35	207	29
Magnesium	Mg	650	1 740	103	150	46	26
Nickel	Ni	1 453	8 860	450	90	68·4	13
Silver	Ag	961	10 490	234	418	16·2	19
Tin	Sn	232	7 300	224	64	114	23
Zinc	Zn	419	7 130	387	111	59·2	26

1. The Structure of Metals and Types of Bonding in Materials

THE STRUCTURE OF THE ATOM

A useful starting point in the study of the structure of metals is to consider the structure of the atom. Although the properties of metals vary widely they possess certain characteristic properties, such as good thermal and electrical conductivity, good malleability and ductility, relatively high melting points and densities, metallic lustre and opaqueness. These and other properties can be explained in terms of atomic structure.

The atom may be regarded as a small-scale solar system consisting of:

(i) A central *Nucleus* which carries a positive charge known as its *Atomic Number*. The nucleus accounts for most of the mass of the atom.

(ii) *Electrons*, each with a negative charge which can be considered as moving in orbits around the nucleus. The number of electrons is equal to the net positive charge on the nucleus.

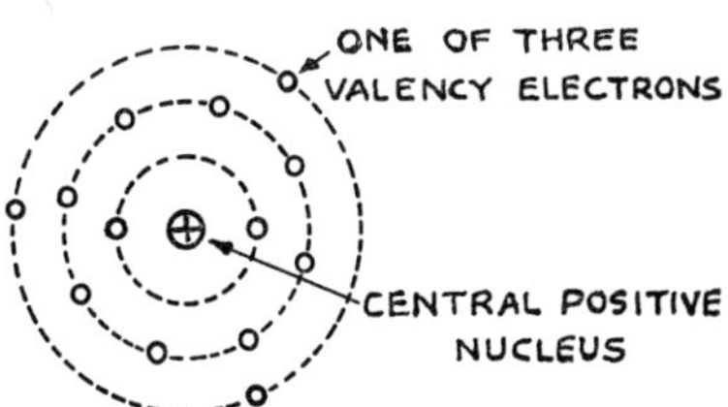

Fig. 1.1. The Structure of the Aluminium Atom (2,8,3)

The structure of the aluminium atom is shown in Fig. 1.1. It will be seen that there are thirteen electrons arranged in three orbits, with two electrons in the first orbit or 'shell', eight in the second, and three in the outer orbit. The atom as a whole is neutral since there are thirteen electrons, each with a negative charge of one, and a positive charge of thirteen on the nucleus. The electrons in the outer shell are referred to as valency electrons and these are very loosely held compared with the other electrons. Metal atoms are capable of loosing these electrons to form positive ions. For example, when the neutral aluminium atom looses three negative electrons it becomes an aluminium ion with a net positive charge of three.

In chemical reactions atoms tend to gain or lose valency electrons in order to obtain a stable grouping of eight electrons in the outer orbit. Metals tend to lose electrons to form positive ions whilst non-metals tend to gain electrons to form negative ions. For example, sodium, with an electronic structure of (2,8,1) can lose its valency electron to form a positive ion, Na^+, with a net positive charge of one. Chlorine, with an electronic structure of (2,8,7), can gain one electron to form a negative ion, Cl^-, with a net negative charge of one.

BONDING IN MATERIALS

The types of bonding in engineering materials may be classified as follows:

1. Ionic Bond
2. Covalent Bond
3. Van der Waal's Forces
4. Metallic Bond

1. Ionic Bond

Ionic crystals are bound together by the electrostatic attraction between the positive and negative ions. Ionic bonds are strong and such crystals are fairly strong with a high melting point. The crystals of sodium chloride, NaCl, are held by ionic bonds between the positive sodium ions and the negative chlorine ions, e.g.

$$Na^+ + Cl^- \longrightarrow Na^+Cl^-$$

2. Covalent Bond

A covalent bond consists of a pair of shared electrons between neighbouring atoms. For example, the carbon atom has four valency electrons and the hydrogen atom has one valency electron. Thus one carbon atom will combine with four hydrogen atoms to form a total of eight valency electrons. This occurs when the compound methane CH_4 is formed. The formula may be represented as follows:

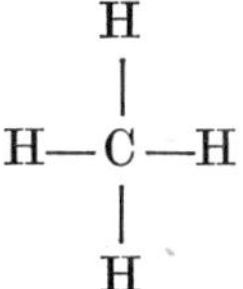

where — represents a pair of shared electrons, i.e. one hydrogen electron and one carbon electron in the pair.

Covalent bonds join the atoms in long chain polymers, e.g. polyethylene $(CH_2)_n$ which can be represented as follows:

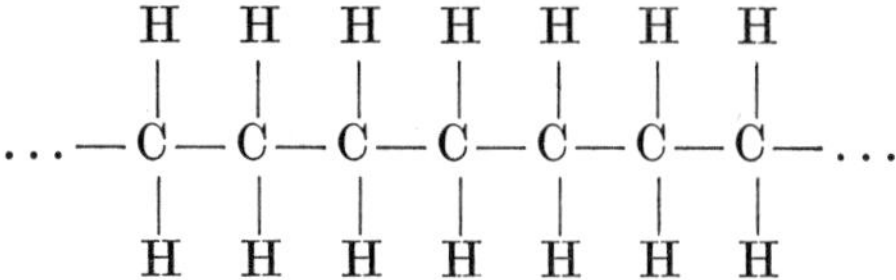

3. Van der Waal's Forces

These are weak surface forces which often exist between the long chain molecules of polymers. The attractive force between the molecules is due entirely to their proximity and not to any chemical bond. Since these forces are weak the chains can slide readily over one another, as they do in thermoplastic materials.

4. Metallic Bond

Metal crystals consist of positive ions arranged on a regular lattice structure immersed in a 'cloud' of free electrons (Fig. 1.2).

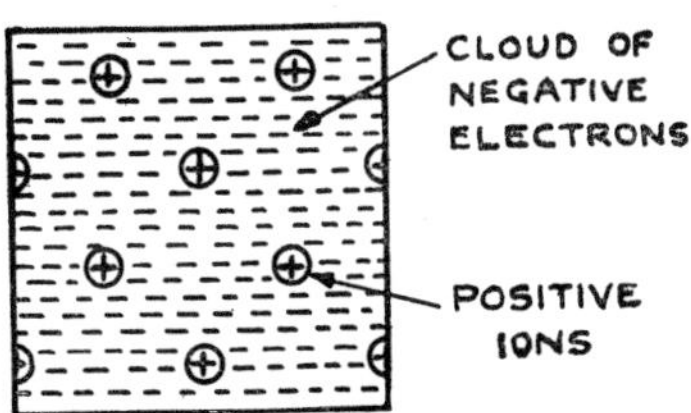

Fig. 1.2. Diagrammatic Representation of Metallic Binding

The electrons of this cloud are not bound to any particular ion but move rapidly through the metal in such a way that an approximately uniform density is maintained. The crystals are held by the electrostatic attraction between the 'cloud' of negative electrons and the positive ions. The free movement of the electrons through the lattice accounts for the good thermal and electrical conductivity of metals. In addition, plastic deformation without fracture can

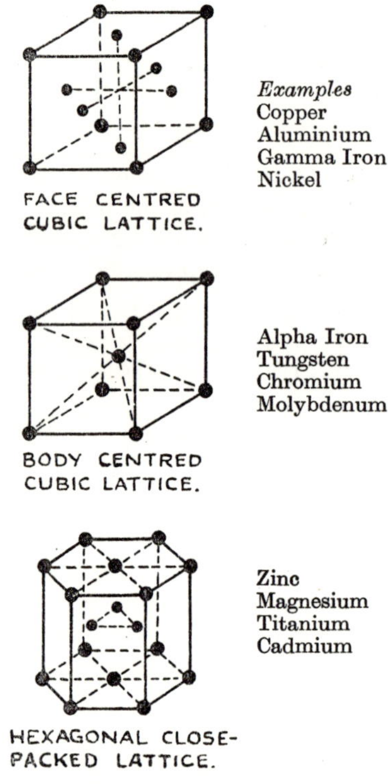

Fig. 1.3. The Main Types of Metallic Space Lattice Arrangements of Atoms

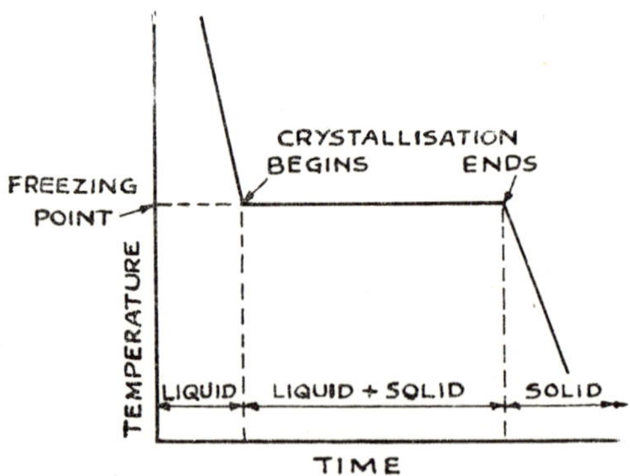

Fig. 1.4. The Cooling Curve for a Pure Metal

occur since as soon as one bond is broken between two ions, another is formed. This accounts for the characteristic malleability and ductility of metals. Since the free electrons in the metal absorb light energy all metals are opaque to transmitted light.

METALLIC SPACE LATTICES

The common metals crystallise in one of three main types of metallic space lattice, namely face-centred cubic, body-centred cubic or hexagonal close-packed. These are shown in Fig. 1.3 and represent the smallest unit of the crystalline metallic structure. It is usual in such diagrams to represent atoms as solid spheres and to ignore the atomic structure.

THE CRYSTALLISATION OF METALS

If a pure metal is melted and the temperature recorded and plotted at various intervals during cooling, the cooling curve for that metal is obtained (Fig. 1.4). The horizontal portion of the curve is due to the evolution of latent heat at the freezing point.

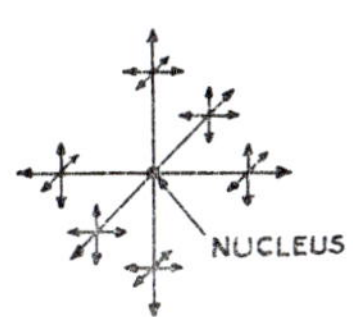

Fig. 1.5. Solidification of Metal Around a Nucleus

Solidification or crystallisation commences by the formation of small 'nuclei' scattered at random in the cooling liquid. At these points a few atoms assume an orderly arrangement to give, for example, the unit cubic structure, and growth occurs in three dimensions as represented in Fig. 1.5.

From the main arms of the crystal, secondary growths occur to give a crystal skeleton known as a dendrite.

Microscopic examination of a pure metal reveals a polygonal grain structure and the formation of these grains from the nuclei by dendritic growth is illustrated in Fig. 1.6.

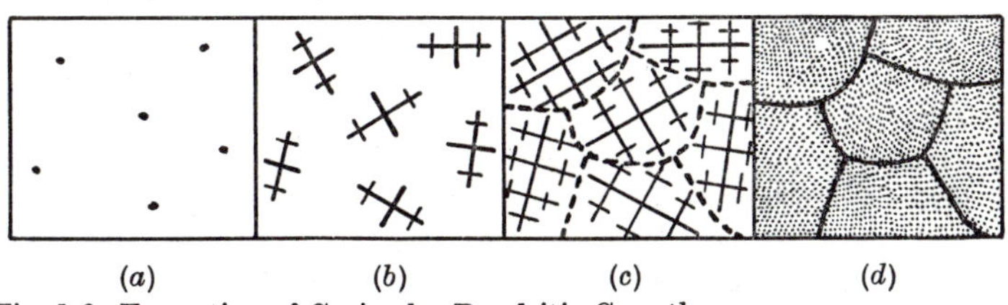

Fig. 1.6. Formation of Grains by Dendritic Growth
(*a*) The formation of nuclei in the cooling liquid.
(*b*) Dendritic crystals grow outwards from nuclei.
(*c*) The dendritic arms meet. Growth outwards is impeded. The contact surface forms grain boundary.
(*d*) The liquid between the arms of the dendrites solidifies giving homogeneous grains with no evidence of dendritic growth.

Each of these grains or crystals is built up of thousands of small unit cubes or cells. In each grain the axes of the cubes all point in the same direction, but this direction varies from one crystal to another, as shown in Fig. 1.7. This effect is known as the orientation of the atoms.

Fig. 1.7. Orientation of Atoms in Each Grain

THE GRAIN BOUNDARY

Since the orientation of the atoms in each grain is different it is obvious that the atoms of the metal at the grain boundary cannot be arranged on a regular space lattice. It is thought that a transition

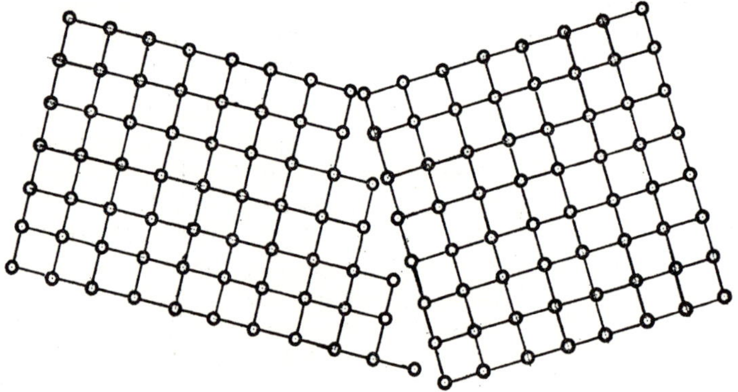

Fig. 1.8. Diagrammatic Representation of Atoms in Adjacent Grains showing Misfit at Grain Boundary

lattice exists at the grain boundary joining the regular cubic lattice of one grain to that of the other. Since the grain boundary structure is different from that of the grains, the properties will also be different.

Reference to Fig. 1.9 will show that at relatively low temperatures the grain boundaries are stronger than the grains, whereas at elevated temperatures the grain boundaries are weaker. It follows that a fine grained structure will give higher strength and hardness at room temperatures, but at higher temperatures a coarse grain would be preferred. These remarks apply to pure metals and solid solution alloys and may not apply to certain alloys with grain boundary impurities. The diagram also explains why, other things being equal, the type of fracture at room temperatures is usually transcrystalline (across the grain) whereas at elevated temperatures fractures may be intercrystalline (along the grain boundaries) (Fig. 1.10).

It is apparent that the properties of a metal will be governed to a large extent by the amount of grain boundary, i.e. the GRAIN SIZE. The control of grain size is therefore very important in the working and heat-treatment of metals and alloys.

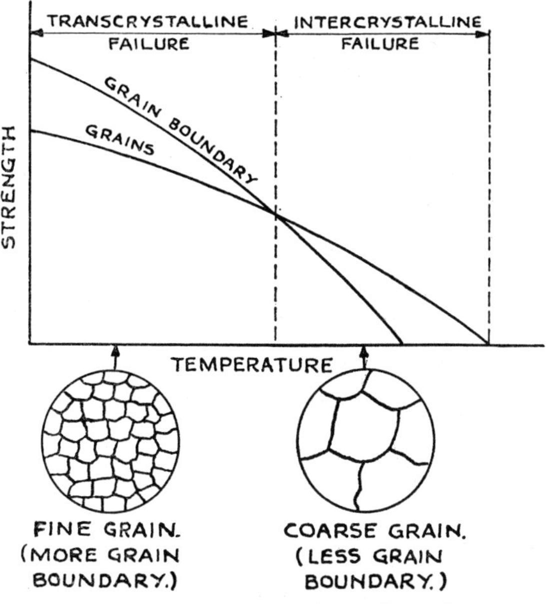

Fig. 1.9. Graph to show Variation of Strength of Grains and Grain Boundaries with Temperature

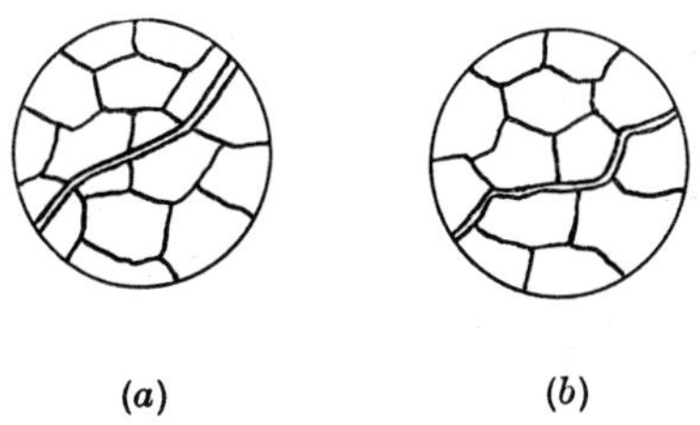

Fig. 1.10. Diagrammatic Representation of (a) Transcrystalline Fracture (b) Intercrystalline Fracture

THE CRYSTAL STRUCTURE OF INGOTS AND CASTINGS

Molten metal from the furnace is usually cast into metal moulds or sand moulds. The rate of cooling of the liquid metal is greater in the former and this increases the number of nuclei formed. Since each nucleus forms a grain, a finer grain size is obtained than in a sand casting.

In addition to the type of mould used the following factors are also important in controlling the grain size of cast metals.

1. Casting Temperature (Fig. 1.11). When the liquid metal makes contact with the cold metal mould a thin layer of chill crystals is formed. Long columnar crystals then grow towards the centre of the ingot from each face. If the casting temperature is too high these crystals will meet to form planes of weakness. However, if the casting temperature is correct, the liquid at the centre of the ingot will have solidified to give equiaxed grains before the columnar crystals can meet.

2. Section Thickness. Thin sections will cool more quickly and consequently a finer grain size is obtained compared with the thicker sections of a casting.

3. Purity of the Metal. In general the purer the metal the coarser the grain.

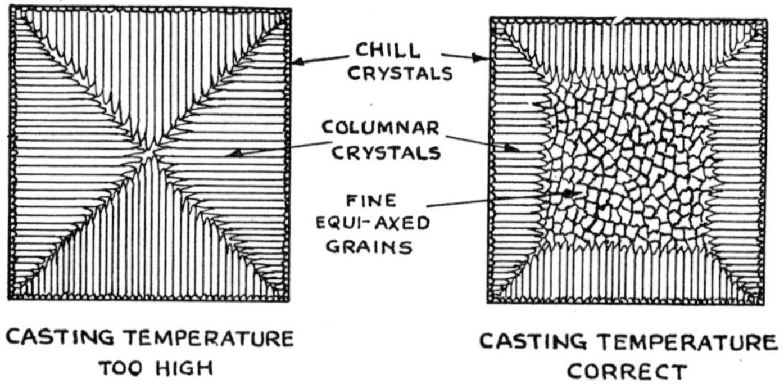

Fig. 1.11. Effect of Casting Temperature on the Structure of Chill Castings

The cast structure with its directional crystal growth is relatively weak, but as indicated later, the structure can be refined by subsequent working operations with consequent improvement in mechanical properties.

2. The Deformation and Annealing of Metals

THE MECHANISM OF PLASTIC DEFORMATION

Most metals are capable of plastic deformation before fracture. If this were not so, it would be impossible to manufacture metal articles by such processes as rolling, drawing and pressing.

Plastic deformation in metal crystals usually occurs by a process of slipping. As a result slip lines, in the form of regularly spaced parallel steps, can be seen on the surface. Slip occurs more easily on the densely packed planes of atoms and early workers likened slip to the sliding of a pack of cards over one another. Deformation of the crystals takes place by shear even though the metal is in tension or compression. Slip occurs when the shear stress on the slip plane reaches a critical value.

The idea that slip occurred by the simultaneous movement of one block of atoms over the other proved satisfactory until theoretical calculations of the strength of metals based on this assumption gave results about a thousand times greater than the observed strengths. This discrepancy between the theoretical and actual cohesive strength of metals has led to the dislocation theory of slip.

DISLOCATION THEORY OF SLIP

The basic idea of this theory is that slip takes place progressively by the movement of what is known as a dislocation along the slip plane. A dislocation is a region of misfit separating the slipped region from the unslipped region. This idea of progressive slip is represented in Fig. 2.2 and is analogous to the motion of a compression wave or the hump of an inch worm.

With this mechanism the dislocation can travel across the crystal with very little applied force. When it has travelled across the crystal the whole plane has slipped one atomic distance. Since this is very

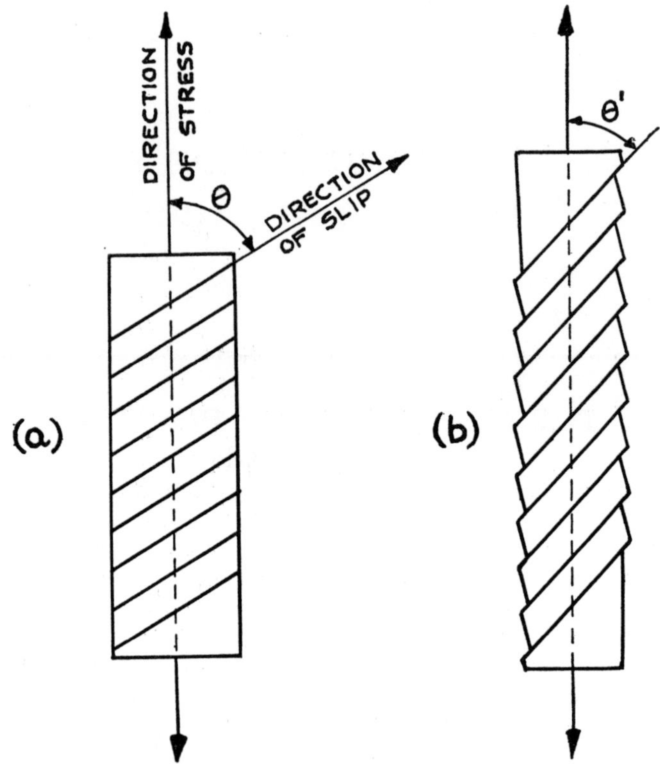

Fig. 2.1. Diagrammatic Representation of Slip in a Metal Crystal
(*a*) Before Slip (*b*) After Slip $\theta > \theta^1$

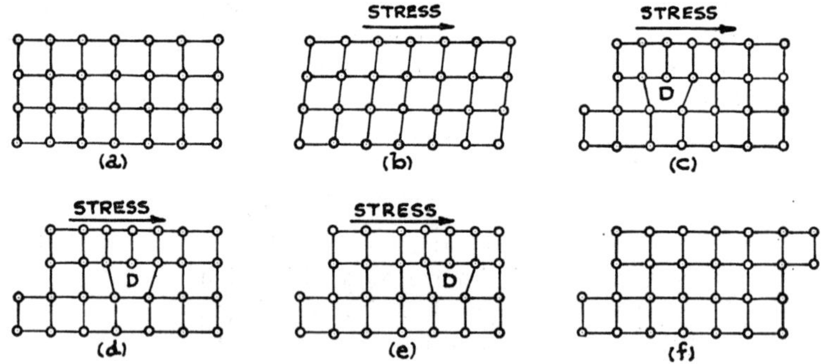

Fig. 2.2. Diagrams to represent progressive movement of a dislocation through a crystal

(*a*) Unstressed crystal lattice
(*b*) Elastic deformation
(*c*) (*d*) (*e*) Plastic deformation. Dislocation represented by letter D moves progressively to the right
(*f*) Plastic deformation complete. Dislocation passes out of the crystal

Redrawn and Adapted from *Theory of Metal Cutting* by Black (McGraw-Hill Book Company Inc.)

small (of the order of 10^{-7} mm), a measurable plastic strain will involve the movement of a large number of dislocations.

DEFORMATION BY TWINNING

Another type of deformation is that of twinning. The formation of a twin band is shown in Fig. 2.3.

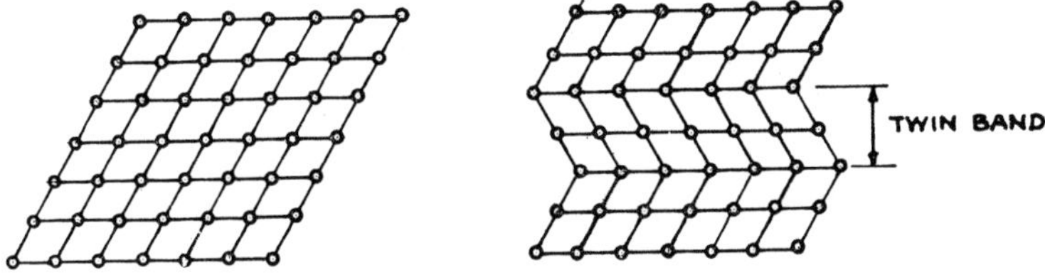

(*a*) Before deformation (*b*) After deformation

Fig. 2.3. Diagram to illustrate Deformation by Twinning

The orientation of a twin band differs from that of the rest of the lattice and is revealed by etching as shown in Fig. 10.3 (*a*). A twinned crystal structure is characteristic of face-centred cubic metals in the cold worked and annealed condition. Twinning occurs by deformation mainly in body-centred cubic and hexagonal close packed metals.

THE COLD-WORKING OF METALS

Cold-working, e.g. rolling, drawing or pressing, is usually carried out on previously hot-worked metals and alloys. It is frequently the finishing stage in production. The effect of cold-working is to break down the crystal structure, elongating the grains in the direction of working (Fig. 2.4).

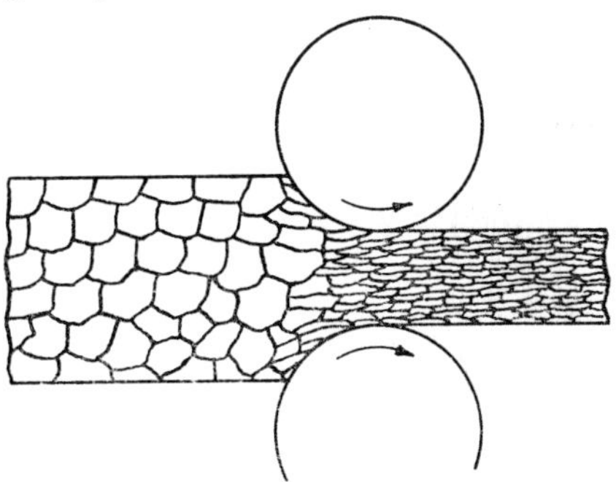

Fig. 2.4. Diagrammatic Representation of Changes occurring during Cold-working

Cold-work destroys the lattice structure with its regular crystal planes along which deformation can readily occur. The metal thus becomes harder and the characteristic ductility is lost. Hardening due to cold-working is referred to as WORK-HARDENING and is the only method available for increasing the hardness of pure metals and many non-ferrous alloys.

The advantages of cold-working over hot-working are:

1. More accurate control of dimensions
2. The production of a clean, bright finish
3. Improvement in yield point, hardness and machinability

THE ANNEALING OF COLD-WORKED METAL

Before further forming processes, e.g. pressing or deep drawing, can be carried out on cold-worked metal it is necessary to soften the material. This is achieved by annealing, which involves reheating the metal, when the following changes occur (Fig. 2.5):

(i) stress relief (ii) recrystallisation (iii) grain growth

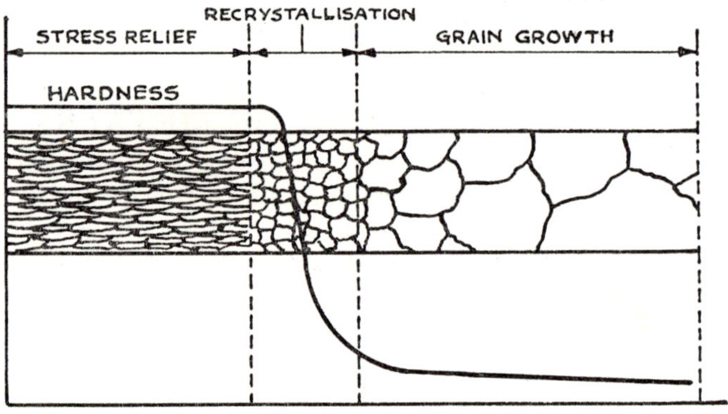

Fig. 2.5. Effect of Annealing Temperature on the Structure and Hardness of Cold-worked Metals

Stress Relief

At relatively low temperatures depending upon the particular metal or alloy the internal stresses induced by cold-working are removed. There is little change in hardness or microstructure during this stage. Stress relief annealing is necessary to avoid stress-corrosion cracking in service.

Recrystallisation

When a certain temperature is reached the distorted grains are replaced by new fine polygonal grains and softening occurs. This phenomenon is known as recrystallisation. The recrystallisation temperature for a particular metal or alloy will depend upon a number of factors, namely:

(a) The degree of prior cold-work

The greater the amount of prior deformation the lower the recrystallisation temperature (Fig. 2.6).

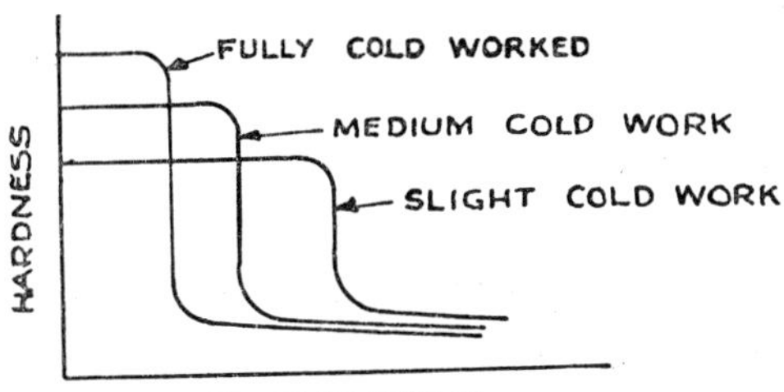

Fig. 2.6. Effect of Cold-working on Softening Temperature

(b) The addition of other elements

Added elements generally raise the recrystallisation temperature. Pure copper recrystallises at 200°C., whilst the addition of 0·25% tin raises the recrystallisation temperature to above 375°C. It follows that alloys generally recrystallise at higher temperatures than the metals from which they are composed.

(c) The annealing time

Increasing time of annealing displaces recrystallisation to a lower temperature.

The approximate recrystallisation temperatures of some commercially pure fully cold-worked metals is shown in Table 2.1.

It will be noticed that certain metals, e.g. zinc, lead, tin, recrystallise at room temperatures; therefore these metals cannot be work-hardened since recrystallisation occurs simultaneously with deformation.

Grain Growth

If the annealing temperature is further increased the recrystallised grains grow in size. The new strain-free grains grow by absorbing

Metal	*Approx. Recrystallisation temperature °C.*	*Melting Point °C.*
Nickel	600	1 455
Iron	450	1 533
Copper	200	1 083
Aluminium	150	660
Magnesium	150	651
Zinc	Room temp.	419
Lead	,, ,,	327
Tin	,, ,,	232

Table 2.1. Table to illustrate melting points and approximate recrystallisation temperatures of some fully cold-worked metals

others which are less stable. The two processes of recrystallisation and grain growth are inseparable but it is more convenient to study them separately.

The factors governing grain growth are as follows:

(a) The annealing temperature and time (Fig. 2.7).

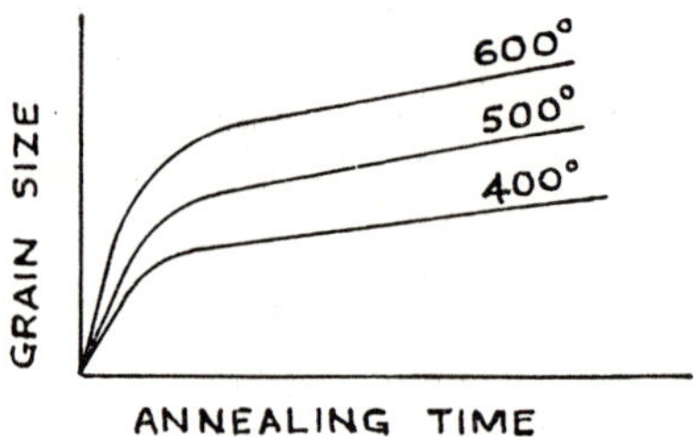

Fig. 2.7. Effect of Annealing Temperature and Time on Grain Size

Reference to Fig. 2.7 indicates that the higher the annealing temperature the coarser the grain size. Grain growth is rapid at first and then becomes much slower.

(b) The degree of previous cold-work

In general the greater the amount of cold-work the finer the grain size after complete recrystallisation. In certain cases, a slight amount of deformation may give rise to an extremely large grain after recrystallisation (Fig. 2.8). This critical amount of deformation due to cold-work varies according to the metal, being about 10%

for iron. In this case deformation is defined as the percentage reduction in thickness or diameter due to cold-working.

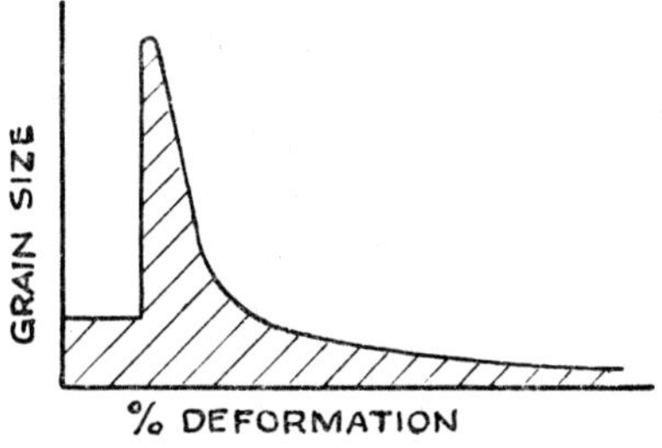

Fig. 2.8. Effect of Degree of Previous Cold-work on Grain Size

(c) The presence of insoluble impurities
These tend to prevent grain growth.

The annealing temperature and time should be closely controlled to avoid grain growth. A coarse grain produces a roughened surface ('orange peel' effect) in subsequent pressing operations. A coarse-grained alloy cannot be remedied by further heat-treatment. Recrystallisation after cold-work is the only method of grain refinement available for most non-ferrous metals and alloys.

HOT-WORKING

The distinction between hot-working and cold-working is that the former takes place above the recrystallisation temperature (or lower critical range for steel). In hot-working, deformation and recrystallisation occur simultaneously (Fig. 2.9) so that the rate of softening is greater than the rate of work-hardening.

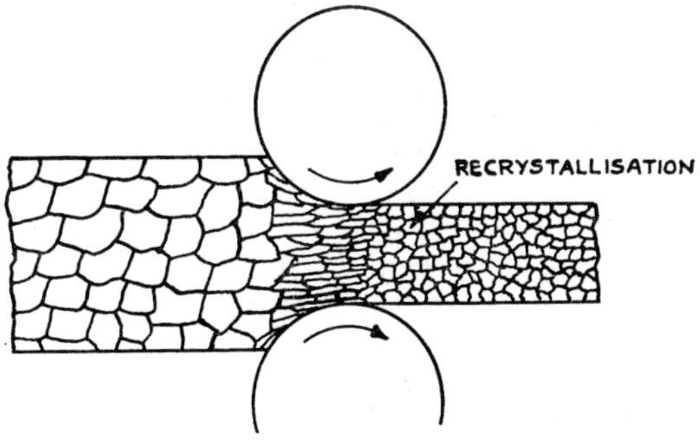

Fig. 2.9. Structural Changes occurring during Hot-working

The important factor in hot-working is the finishing temperature. This is apparent from Fig. 2.10. Hot-working should be finished at a temperature just above the recrystallisation temperature so that a fine grain size is obtained. If the finishing temperature is too high then grain growth will occur whilst the metal is cooling above the recrystallisation temperature. If on the other hand the finishing temperature is too low then work-hardening will result.

Far less power is required to deform metals at high temperatures since the metal is usually softer and more plastic. Hot-working is therefore more economical. The chief processes of hot-working are hot-rolling, forging and extrusion. In general, simple shapes, e.g. sheet, plate, rod, are usually hot-rolled, whilst the more complicated shapes are forged. Non-ferrous sections and tubes are usually made by extrusion.

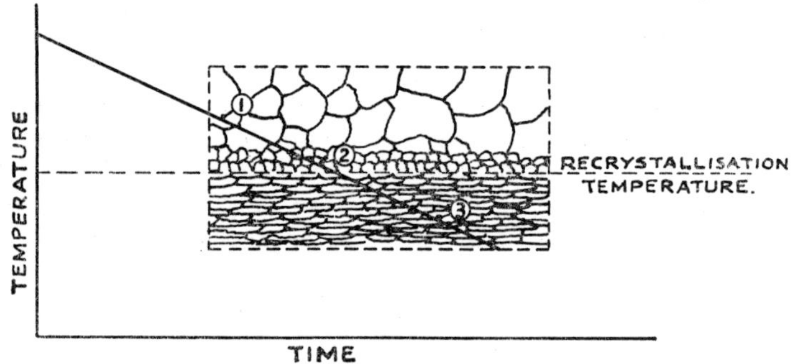

Fig. 2.10. Effect of Finishing Temperature on the Structure of Hot-worked Metal

1. Finishing temperature too high		Coarse grain
2. „ „	correct	Fine grain
3. „ „	too low	Distorted grain Work hardening

EFFECTS OF HOT-WORKING

1. The ingot structure, with its coarse columnar crystals, is destroyed.

2. If the finishing temperature is correct, grain refinement is obtained.

3. The non-metallic inclusions are elongated into fibres, thus indicating the flow of the metal.

4. In certain cases, e.g. with rimming steels (page 31), internal blowholes may be welded up.

5. Strength, ductility and toughness are improved but directional

properties are produced. The improvement is greater in the direction of working than in the transverse direction (Fig. 2.11(*a*)). Table 2.2.

In forgings, the 'fibre' or flow lines should follow the contour of the section (Fig. 2.11(*b*)). Forgings are, therefore, superior in mechanical properties to similar shapes machined from hot-rolled material (Fig. 2.11(*c*)).

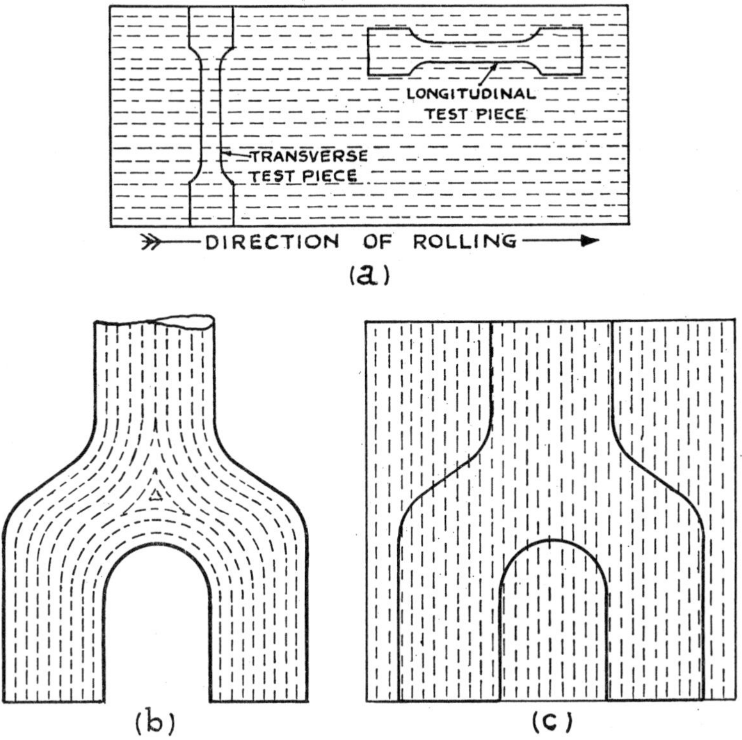

Fig. 2.11. Effect of Direction of Working or 'Fibre' on Mechanical Properties

Condition	*TS* *N/mm²*	*El*	*Izod* *(Joules)*
Cast	420	13	7
Hot rolled (longitudinal)	460	34	41
Hot rolled (transverse)	430	18	16

Table 2.2. Effect of hot working on the properties of a mild steel ingot

3. Thermal Equilibrium Diagrams

Pure metals are often too soft and weak for many commercial applications and consequently alloys are more widely used as engineering materials. For example, steel (an alloy of iron and carbon) is stronger than pure iron and can also be heat-treated to produce desirable mechanical properties. Alloy systems can best be studied with reference to constitutional or equilibrium diagrams. These diagrams are really temperature-composition diagrams which indicate the structural changes that take place during the heating and cooling of an alloy. Although they only refer to equilibrium conditions, which are rarely obtained in practice, they form a useful basis for the study of the treatment and properties of alloys.

Alloys containing two metals are referred to as binary alloys. Even when more than two metals are present, much useful information can be obtained by the study of the binary diagram for the two principal metals. The constituent metals in most commercial binary alloys are soluble in each other in the liquid state. Assuming that they do not combine to form a compound, binary alloys can be divided into three types, namely:

(*a*) The two metals are completely insoluble in each other in the solid state (*simple eutectic type*).

(*b*) The two metals are completely soluble in each other in the solid state (*solid-solution type*).

(*c*) The two metals are partially soluble in each other in the solid state (*combination type*).

Each of these types gives rise to a characteristic equilibrium diagram which may be recognised at a glance. These are illustrated in Fig. 3.1.

In certain cases a peritectic reaction may occur during solidification. The meaning of this and other terms will be explained later,

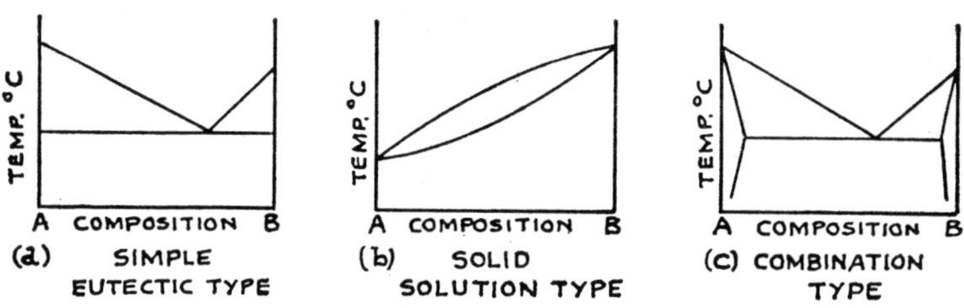

Fig. 3.1. Characteristic Types of Binary Equilibrium Diagrams for Alloys of the Metals A and B

but here again a characteristic type of diagram may be recognised (Fig. 3.2).

The equilibrium diagrams for the chief engineering alloys, e.g. steels (Fe-C), brasses (Cu-Zn), bronzes (Cu-Sn), etc., are all complex diagrams, but they can be interpreted on the basis of the simple

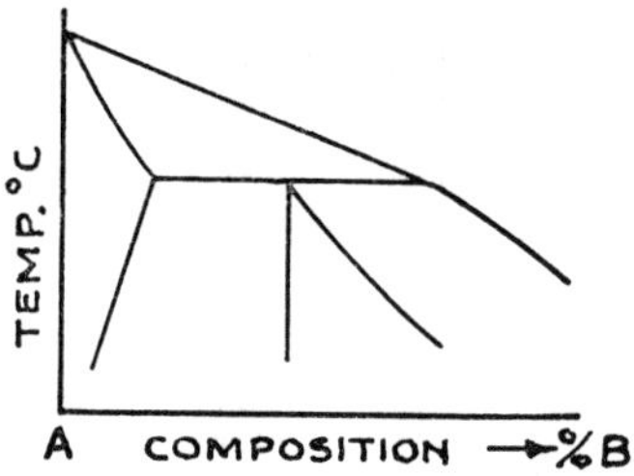

Fig. 3.2. Part of Equilibrium Diagram of a Binary Series of Alloys involving a Peritectic Reaction

types mentioned above, viz. eutectic, solid solution and peritectic. It is therefore essential to understand these basic diagrams before proceeding to study commercial alloy systems.

COOLING CURVES FOR BINARY ALLOYS

The cooling curve for a pure metal is shown in Fig. 1.4. It will be noticed that freezing takes place at a constant temperature. However, with the exception of alloys of exact eutectic composition, alloys freeze over a range of temperature, as indicated in Fig. 3.3.

The points A and B denote the temperatures corresponding to the beginning and end of freezing, and are known as the first and second arrest points respectively. In Fig. 3.3(*a*) it will be observed that solidification finally occurs at a constant temperature and this is true for all alloys which form a eutectic.

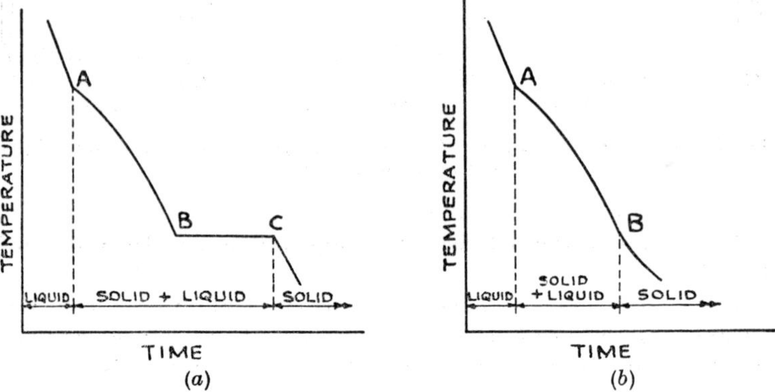

Fig. 3.3. (*a*) Cooling Curve of Eutectic-type Binary Alloy which is not of exact Eutectic Composition
(*b*) Cooling Curve of Binary Alloy forming a Single Solid Solution

THE CONSTRUCTION OF THERMAL EQUILIBRIUM DIAGRAMS

e.g. The Simple Eutectic Type

Equilibrium diagrams may be constructed using the information supplied from the cooling curves of a number of alloys of the two metals concerned.

The upper diagram in Fig. 3.4 represents a series of cooling curves obtained using various alloys of the two metals A and B. The first curve refers to pure metal A, the other curves to alloys containing (80%A 20%B), (60%A 40%B), (40%A 60%B), (20%A 80%B), while the last curve is that for pure metal B. By joining the first arrest points (represented by circles) we obtain the liquidus curves, above which the alloys are entirely liquid. By joining the second arrest points (represented by crosses) we obtain the solidus line, below which the alloys are entirely solid. Between the solidus and the liquidus the alloy is a paste of part solid and part liquid.

It is apparent from Fig. 3.4 that by increasing the proportion of metal B to metal A the freezing point has been lowered as shown by curve AC. Similarly curve BC shows the effect of increasing the proportion of metal A to metal B. At C, the point of intersection, we have an alloy with the lowest freezing point in the series. This alloy is known as the *eutectic*, the word being derived from the Greek meaning 'easy melting'. At the eutectic temperature both metal A and metal B will crystallise simultaneously at a constant temperature. Eutectics usually possess a laminated structure, consisting of alternate laminations of the two constituents. In this example we

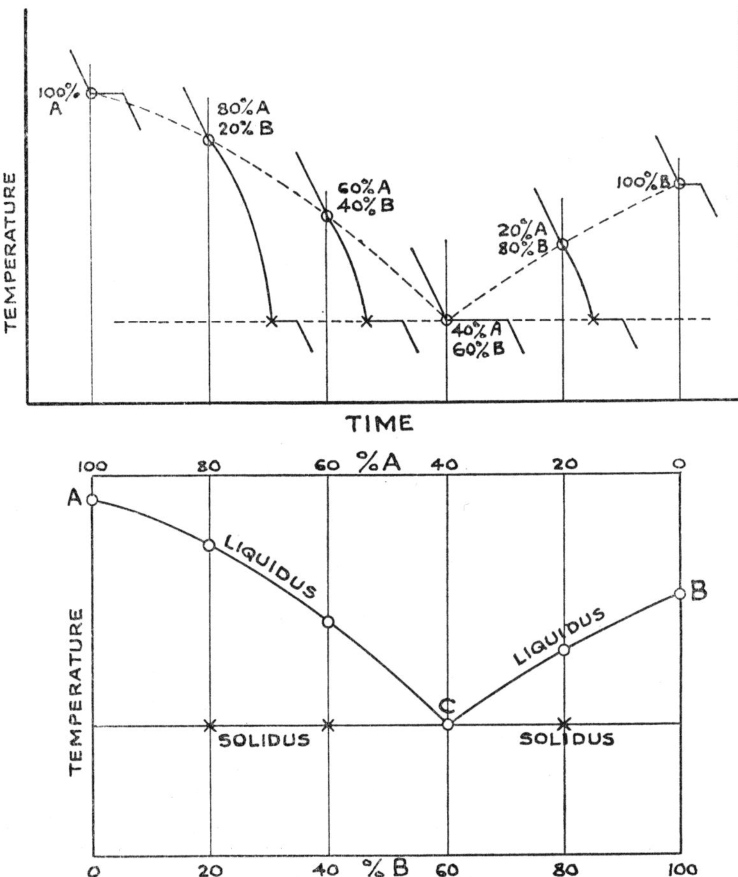

Fig. 3.4. Construction of a Thermal Equilibrium Diagram from Cooling Curves

have a eutectic of two pure metals, but eutectics may consist of solid solutions, or chemical compounds. A comprehensive definition of a eutectic is as follows:

A eutectic consists of two or more solid phases produced by solidification at constant temperature. It is the alloy of the lowest freezing point (melting point) in the series.

THE INTERPRETATION OF THE SIMPLE EUTECTIC DIAGRAM

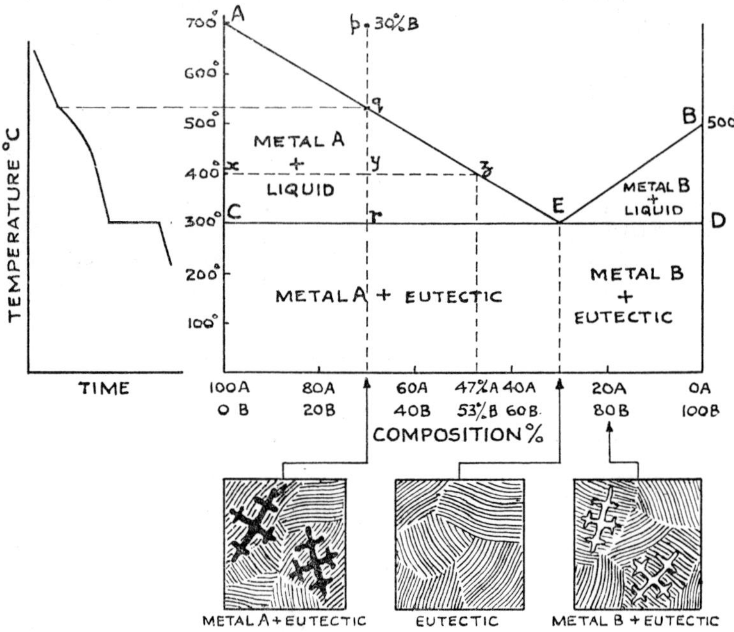

Fig. 3.5. Thermal Equilibrium Diagram of the Simple Eutectic Type

Let us consider a hypothetical simple eutectic diagram of two metals A and B. Metal A melts at 700°C. and metal B at 500°C. They form a eutectic containing 70%B 30%A which melts at 300°C.

Consider the cooling of an alloy containing 30%B. The alloy contains more of metal A than required to form a eutectic. Upon reaching point q on the liquidus (approximately 525°C.) crystals of metal A will be formed. As the temperature decreases more crystals of metal A are deposited and the liquid becomes progressively richer in metal B, as represented by the liquidus qE.

At 400°C. the alloy will consist of solid metal A plus liquid of composition z (53%B 47%A). The relative weights of solid and liquid are given by the relative lengths of the lines yz and xy.

$$\frac{\text{Weight of solid}}{\text{Weight of liquid}} = \frac{yz}{xy} = \frac{23}{30}$$

Upon reaching point r (300°C.) the liquid has attained the eutectic composition E (70%B 30%A). At this temperature, both metals A and B will crystallise simultaneously to form the eutectic structure. No further changes will occur upon cooling to room temperatures. The final microstructure will therefore consist of dendrites of metal A plus eutectic. All the alloys containing up to 70%B will consist of these two phases, but the proportion of eutectic will increase with increasing content of metal B.

Alloys with greater than 70%B will commence to solidify by depositing metal B. The residual liquid will become progressively richer in metal A as the temperature falls until it reaches a composition of 30%A 70%B at 300°C. At this temperature both metals will crystallise simultaneously as eutectic.

The alloy of composition E will solidify entirely as eutectic at a constant temperature (300°C.).

It will be apparent that microscopic examination can be used to estimate the composition of an alloy. For example if examination revealed approximately equal proportions of metal A and eutectic the composition would be mid-way between 0 and 70%B, giving a composition of 35%B 65%A.

Examples of alloy systems which give rise to simple eutectic diagrams are the bismuth-cadmium and zinc-tin alloys. However, most metals usually have a slight solid solubility. Eutectic alloys are widely used for soldering and brazing and for casting alloys.

EUTECTOID

At this stage it is convenient to define the term 'eutectoid' which is closely related to a eutectic. A eutectoid consists of two or more solid phases produced by the breakdown of a solid solution at constant temperature, whereas a eutectic is formed by solidification. Examples of eutectoid formation will be discussed when considering the steels, tin bronzes, and aluminium bronzes.

SOLID-SOLUTION ALLOYS

In certain alloys the complete solubility that exists in the liquid state persists after solidification. The solid alloy, known as a solid solution, consists of one kind of crystal lattice structure in which both metals are present. However, if the alloy is examined microscopically it is impossible to trace the two constituent metals, since a single-phase structure exists.

There are two types of solid-solution alloy, namely substitutional and interstitial.

(a) Substitutional Solid Solutions

The atoms of the added metal can be substituted for those of the parent metal on the lattice. In such cases the metals must possess nearly equal atomic diameters. Copper and nickel are mutually soluble in all proportions to form substitutional solid solutions (Fig. 3.6).

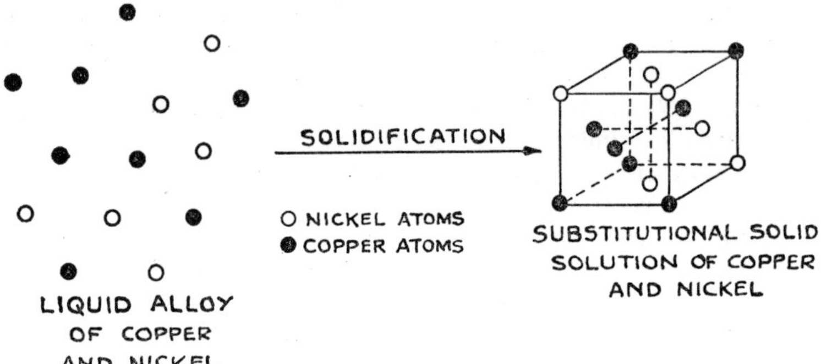

Fig. 3.6. Formation of a Substitutional Solid-solution Alloy

(b) Interstitial Solid Solutions

The atoms of the added elements enter the interstices of the parent lattice. In other words, they fit into the spaces between the atoms of the parent metal. This is of less common occurrence and is only possible if the atoms of the added element are small compared with those of the parent metal. A good example is that of carbon in iron to form the various steel solid solutions (Fig. 3.7).

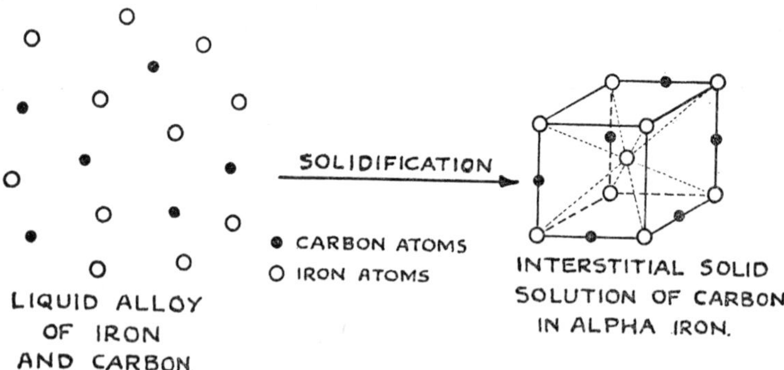

Fig. 3.7. Formation of an Interstitial Solid-solution Alloy

In certain alloys containing three metals, known as ternary alloys, both types of solid solution may co-exist. For example, in austenitic manganese steel there is a substitutional solid solution of manganese and iron and also an interstitial solid solution of carbon in iron.

THE SOLID-SOLUTION DIAGRAM

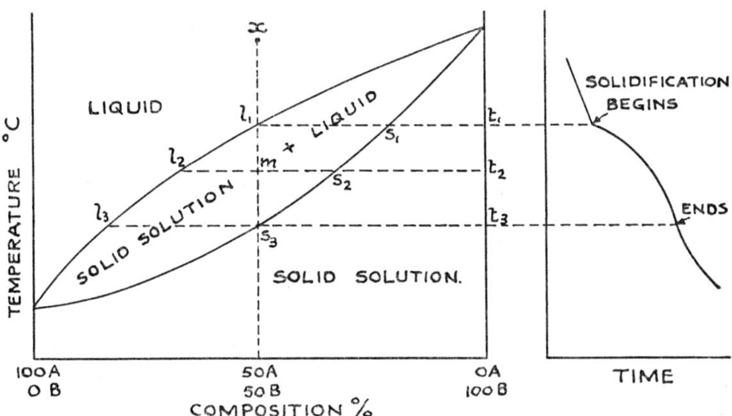

Fig. 3.8. Thermal Equilibrium Diagram of the Solid-solution Type

This diagram may be constructed in a manner similar to that used for the simple eutectic type, namely by joining the first and second arrest points obtained from a series of cooling curves of alloys in the system.

A typical solid solution type of diagram is shown in Fig. 3.8. The upper curve is the liquidus and the lower curve the solidus.

Consider the cooling of an alloy of composition x, containing equal amounts of the two metals A and B. Solidification commences at t_1 when a solid solution of composition s_1 (richer in metal B than 50%) is deposited. Solidification proceeds by the absorption of metal A from the liquid which diffuses throughout the solid. Hence as the temperature falls from t_1 to t_3, the solid solution changes its composition along the solidus s_1 s_3 whilst the liquid changes its composition along the liquidus l_1 l_3. At t_3 the alloy is completely solid and consists of uniform grains of solid solution of composition s_3, the last drop of liquid having the composition l_3.

At some intermediate temperature t_2 we have a solid solution of composition s_2 in equilibrium with a liquid solution of composition l_2. The relative weights of solid and liquid are given by the relative lengths of the lines l_2m and s_2m.

$$\frac{\text{Weight of solid solution } s_2}{\text{Weight of liquid solution } l_2} = \frac{l_2m}{s_2m}$$

Effect of Cooling Rate

It has been seen that the initial dendritic deposits are richer in metal B whilst the last liquid to solidify was richer in metal A. With slow cooling rates diffusion has time to occur producing a structure consisting of a single constituent or phase, similar to that of a pure metal. However, in practice, the cooling rate is too rapid to allow diffusion to occur and what is known as a CORED structure results (Fig. 3.9). Coring can be eliminated by annealing, when diffusion of

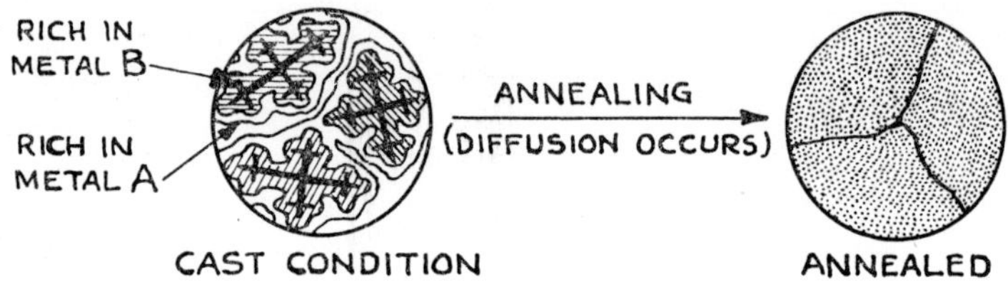

Fig. 3.9. Effect of Annealing on the Microstructure of a Cast Solid-solution Alloy

the two metals occurs. Examples of binary solid-solution alloys are copper-nickel, bismuth-antimony and gold-silver.

THE COMBINATION-TYPE DIAGRAM

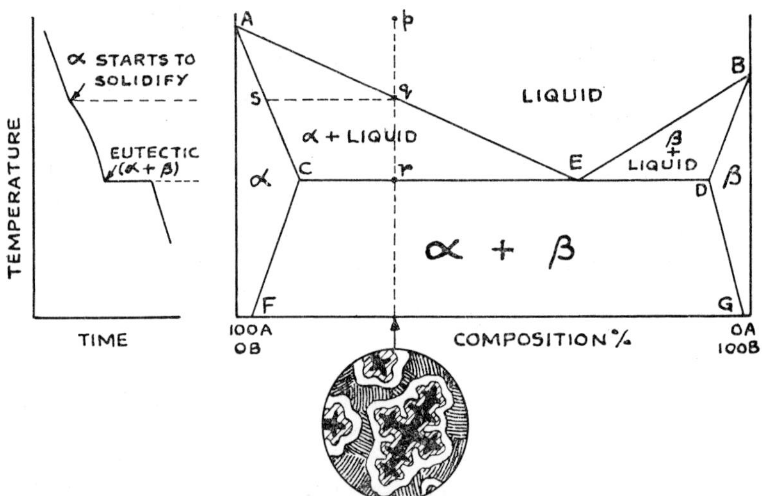

Fig. 3.10. Equilibrium Diagram for Binary Series of Alloys in which two Metals are partially Soluble in each other in the Solid State

This type of equilibrium diagram (Fig. 3.10) is really a combination of the two previous types. The liquidus line AEB is similar to that of the simple eutectic diagram. The solidus line in this diagram is ACEDB and contains two parts, namely, AC and BD which are similar to the solidus line on the solid solution diagram.

The solid solubility of B in A and A in B increases with temperature, as shown by the lines FC and GD respectively. The maximum solubility is reached at the eutectic temperature, when the compositions of the two solid solutions α and β are denoted by the points C and D respectively. In this case the eutectic of composition E consists of the two solid solutions α and β.

Consider the cooling of an alloy of composition *p*. Solidification commences on reaching point *q* on the liquidus line when solid solution α of composition *s* is deposited. When the temperature falls to that of the eutectic, the alloy will now consist of α of composition C and liquid of composition E. The liquid will solidify completely to form a eutectic of the two solid solutions α and β in the proportions $\alpha:\beta = ED:CE$. Upon further cooling the α and β in the eutectic will become poorer in metals B and A respectively. Their compositions at room temperature are represented by F and G respectively. The primary dendrites of α solid solution will also become poorer in metal B, changing in composition along CF, and this will precipitate a small amount of β solid solution, which will be associated with the eutectic.

Examples of binary alloy systems of this type are the lead-tin, copper-silver and bismuth-tin alloys.

PERITECTIC REACTIONS

During the freezing of certain alloys the solid already deposited may react with the residual liquid to form another solid solution or compound of a composition intermediate between that of the first solid and the liquid. This reaction occurs at constant temperature, and the alloy is said to undergo a peritectic transformation.

A peritectic reaction gives rise to a characteristic type of diagram as shown in Fig. 3.11. Consider the cooling of an alloy of composition *p*. Solidification commences at t_1 when crystals of solid solution α of composition *s* are deposited. Throughout the temperature range t_1–t_3 the composition of the α varies along the solidus line AB whilst the liquid varies in composition along the liquidus line AD. At t_3 there exists α of composition B in equilibrium with liquid of composition D in the ratio qD:qB. The peritectic reaction occurs at this temperature when α solid solution of composition B reacts with liquid of composition D to form a new solid solution β of composition C. Since the original liquid is not rich enough in metal B to form entirely β solid solution, some α solid solution remains unchanged. The alloy solidifies at this temperature leaving a structure of α and β of composition B and C respectively. Although the two solid solutions vary in composition on cooling, the general structure still remains the same.

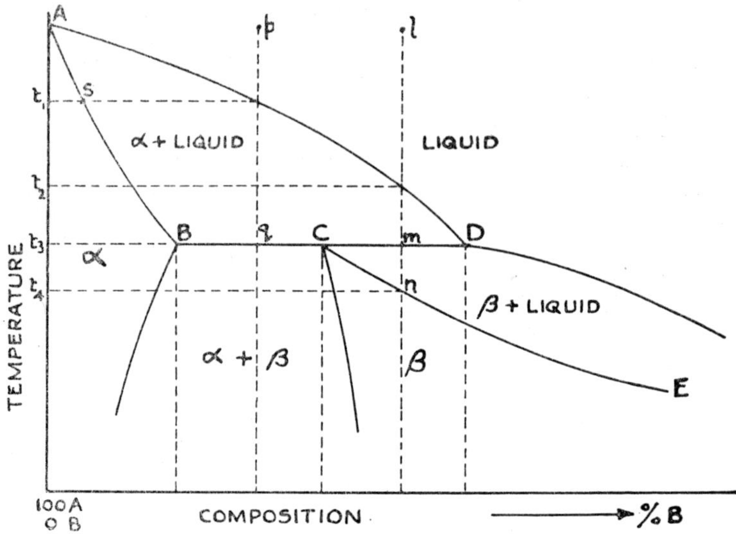

Fig. 3.11. Part of a Binary-alloy Diagram in which a Solid Solution is formed by a Peritectic Reaction

The alloy of composition l contains more of metal B than that required to form the β solid solution. This alloy begins to solidify at t_2, and at t_3 will consist of solid solution α of composition B and liquid solution of composition D in the ratio of mD:mB. The peritectic reaction occurs at this temperature and will proceed until all the α solid solution has transformed, leaving β solid solution + liquid. Upon further cooling the β changes in composition along the solidus CE, being completely solid at temperature t_4 when it consists of uniform crystals of composition l.

Examples of binary systems involving a peritectic reaction are iron-carbon, copper-zinc and copper-tin.

INTERMETALLIC COMPOUNDS

In some alloy systems the two metals may enter into definite chemical combination. The compound produced, known as an *intermetallic compound*, is apparent under the microscope, frequently as definite rectangular particles. The existence of a compound which has a definite melting point is indicated by a maximum in the liquidus curve (Fig. 3.12).

The diagram may be considered as two eutectic diagrams placed together. In the example shown, the compound and each metal are completely insoluble in the solid state. Examples of alloy systems in which an intermetallic compound is formed are magnesium-zinc and magnesium-tin.

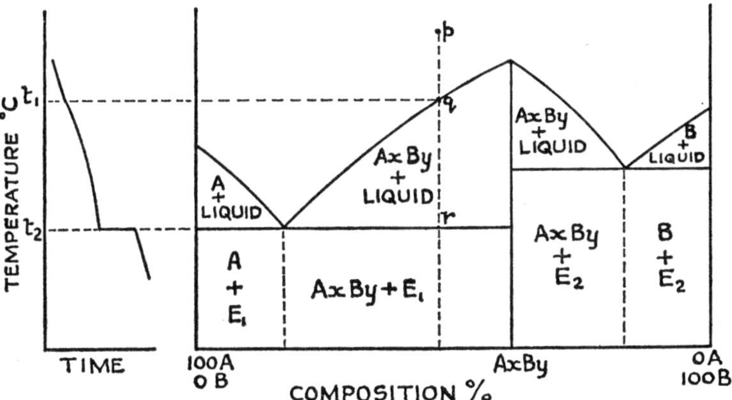

Fig. 3.12. Binary Equilibrium Diagram showing presence of Intermetallic Compound
E_1 = Eutectic of (Ax By + A)
E_2 = Eutectic of (Ax By + B)

Consider the cooling of an alloy of composition *p*. Solidification commences at temperature t_1, represented by point *q* on the liquidus, when crystals of the compound A*x* B*y* are deposited. Upon further cooling the liquid becomes progressively richer in metal A until at temperature t_2, represented by point *r* on the solidus, it has reached the eutectic composition E. At this temperature, both metal A and the compound A*x* B*y* solidify simultaneously as a eutectic at constant temperature. The final structure will therefore consist of primary crystals of the compound A*x* B*y* in a eutectic matrix.

When present as rectangular particles, intermetallic compounds are hard, e.g. SbSn cuboids in white metal bearings and cementite, Fe_3C, in cast irons. However, in the form of fine precipitated particles such compounds can have a strengthening effect, e.g. fine $CuAl_2$ particles in age-hardened Duralumin.

4. Plain Carbon Steels

BRIEF OUTLINE OF STEEL MAKING

Steel making involves the removal, by oxidation, of the impurities from pig iron or a mixture of pig iron and steel scrap. Pig iron contains 3–4% carbon, together with smaller amounts of manganese, silicon, sulphur, and phosphorus. These elements make the iron weak and brittle and their partial removal is necessary to produce a stronger and more ductile product for commercial use. The chief processes available are the open-hearth and the Bessemer. The electric-arc process is also used for making high-grade alloy steels, usually from a charge of scrap steel, although a certain amount of refining is involved.

Steel-making processes are usually classified as either 'acid' or 'basic'. These terms refer to the type of furnace lining and consequently the nature of the slag. In the acid process the furnace lining consists usually of silica. However, in order to remove phosphorus from pig iron it is necessary to add lime, resulting in a basic slag. Because this slag reacts with the acid silica lining, acid processes are unsuitable for treating the pig irons produced from high phosphorus ores. Such pig irons can only be treated in a basic process using magnesite or dolomite as furnace lining. Magnesite consists of magnesium carbonate $MgCO_3$, whereas dolomite is essentially a mixture of the carbonates of calcium and magnesium, $CaCO_3$, $MgCO_3$. At one time basic steel was regarded as inferior to acid steel, but with present techniques basic steel is just as good. A comparison of the two types of processes is shown in Table 4.1.

DEOXIDATION

At the end of the steel-making process some iron will be in the oxidised condition and deoxidation is necessary before pouring the steel into ingot moulds. Deoxidisers, such as ferro-manganese, ferro-silicon and aluminium, are added at the end of the process to remove this soluble iron oxide. These additions have a strong affinity for oxygen and form insoluble oxides of manganese, silicon or aluminium. These insoluble oxides normally enter the slag, but if they do not they will form non-metallic inclusions in the steel.

	Acid	*Basic*
Refractory lining	Silica	Magnesite
Composition of pig iron used	High silicon Low phosphorus	High phosphorus Low silicon
Slag	Acid (high silica)	Basic (high lime)
Elements removed by oxidation	C, Si, Mn	C, Si, Mn, P, S

Table 4.1. Comparison of Acid and Basic Steel-making Processes

Normally such inclusions are not too detrimental. If they are soft they can be readily elongated in the direction of working, giving rise to 'fibre' and directional properties (Fig. 2.11). However, hard inclusions, such as the oxide of aluminium (Al_2O_3), are not elongated and act as points of stress concentration. These inclusions give rise to low Izod impact test values, corrosion fatigue, poor machinability and excessive tool wear. They also cause increased abrasion of dies in wire drawing, as well as causing the steel wire to have a very poor surface.

KILLED AND RIMMING STEELS

Steels that have been completely deoxidised are referred to as 'killed' steels. Such steels are free from blow holes and are characterised by 'piping', a term used to describe the normal solidification shrinkage cavity. This can be minimised by using a wide-end up mould together with a refractory 'hot top' or feeder head (Fig.4.1). Rimming steels are cast in an oxidised condition, deoxidation not being complete. The control of deoxidation is very important in the manufacture of rimming steels. A thick rim of pure metal solidifies and the residual liquid is enriched sufficiently in carbon to react with the oxide.

$$FeO + C = Fe + CO$$

The carbon monoxide, CO, evolved causes a risen surface which counteracts the piping and gives rise to a core which contains numerous blow holes (Fig. 4.1(*b*)). Rimming steels are usually low

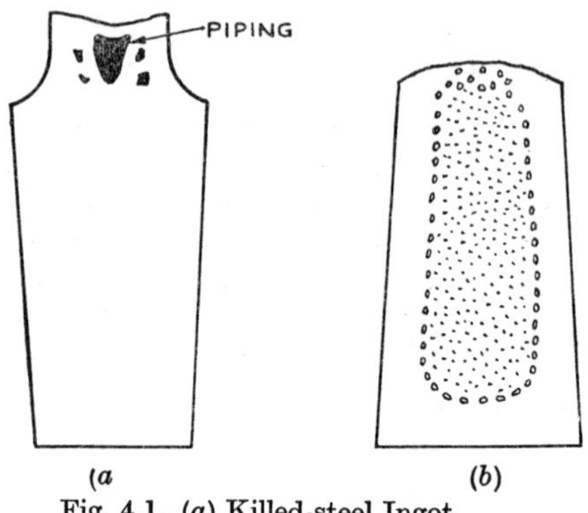

Fig. 4.1. (*a*) Killed-steel Ingot
(*b*) Rimming-steel Ingot

carbon steels, especially suitable for sheet and plate, due to their pure rim. Since the carbon content is low the internal blow holes are welded up during subsequent hot rolling.

STRUCTURE OF PLAIN CARBON STEELS

Allotropy of Pure Iron

Before studying the structure of plain carbon steels it is essential to understand the structural changes that occur during the heating and cooling of pure iron itself. Pure iron exhibits allotropy, which may be defined as the phenomenon of an element existing in more than one physical form. This can be seen by plotting the volume of the unit cell of pure iron against the temperature (Fig. 4.2). A crystal structure exists in the temperature range 937°–1 400°C different from that at other temperatures. This is the allotropic form known as γ iron and possesses a face-centred cubic lattice. At the other temperatures a body-centred cubic lattice exists, this being the crystal structure of α, β and δ iron. The change points (or critical points) denoting the changes α–β, β–γ, γ–δ, are referred to as the A_2, A_3 and A_4 points respectively. The A_2 change point at 769°C. is the temperature at which iron loses its magnetism and is not important in heat-treatment work. In steels there is an additional change point, known as the A_1 point, which is associated with the formation of the eutectoid pearlite.

These structural changes in the solid state involve the evolution of heat and may be studied with reference to an *inverse-rate* cooling curve. Such curves are more sensitive than direct cooling curves in detecting the smaller heat changes that occur in solid-state transformations. Inverse-rate curves are obtained by plotting the

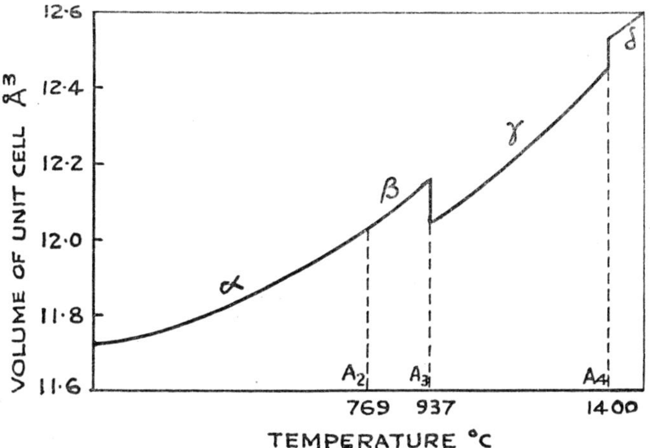

Fig. 4.2. Change in Volume of Unit Cell of Pure Iron with Temperature

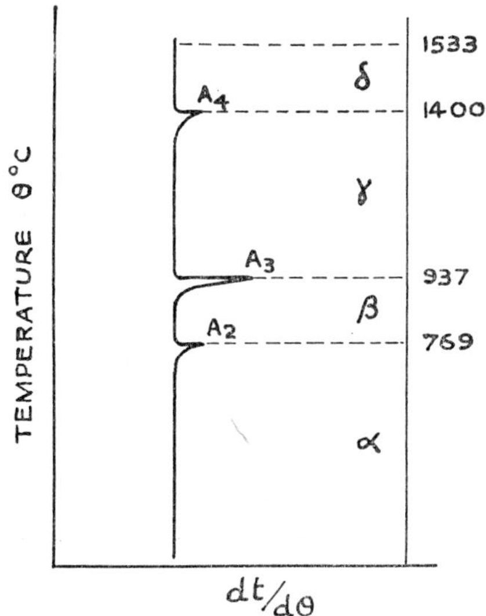

Fig. 4.3. Inverse-rate Cooling Curve for Pure Iron

temperature θ against $\frac{dt}{d\theta}$ where t represents the time. The thermal effect accompanying the phase change is prominent as a peak on the curve. The temperatures given for these various change points are those obtained upon heating the iron. Upon cooling it is noticed that they occur at somewhat lower temperatures. The change points obtained upon heating and cooling are referred to as Ac and Ar points respectively. For example, the Ac_1 and Ar_1 points occur at 730°C. and 695°C. respectively, whilst the Ac_3 and Ar_3 occur at 937°C and 910°C. respectively.

THE 'STEEL PORTION' OF THE IRON-CARBON DIAGRAM

By joining the upper critical points (A_3) and the lower critical points (A_1) obtained from a series of inverse-rate cooling curves using steels of varying carbon contents, the steel portion of the iron-carbon diagram may be constructed (Fig. 4.4). For the purpose of this discussion the term 'steel portion' refers to that part of the iron-carbon diagram of use in the heat treatment of plain carbon steels. It will be noticed that the addition of carbon lowers the A_3 point until at 0·83% carbon it merges with the A_1 point. This is the eutectoid point at 695°C. and is associated with the formation of the structure known as pearlite.

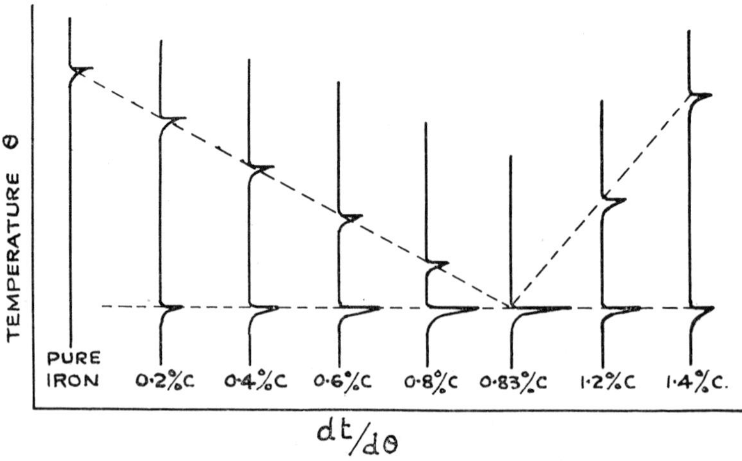

Fig. 4.4. Construction of 'Steel Portion' of Iron-carbon Diagram from Inverse-rate Cooling Curves

The 'steel portion' of the iron-carbon diagram is shown in Fig. 4.5.

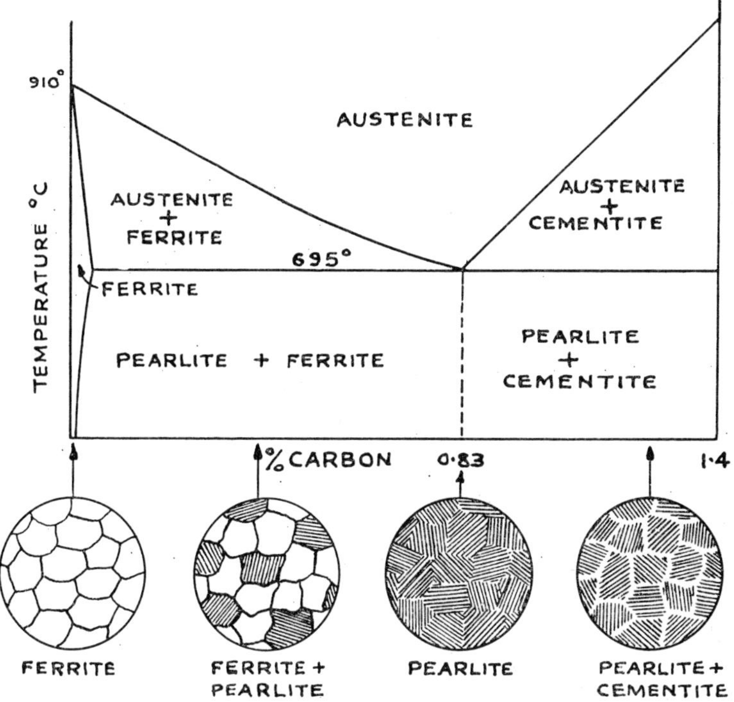

Fig. 4.5. The 'Steel Portion' of the Iron-carbon Diagram

Those steels with less than 0·83% carbon are referred to as hypo-eutectoid steels, whilst those with more than 0·83% are known as hyper-eutectoid steels. The various microconstituents present in plain carbon steels may be defined as follows:

Ferrite

A solid solution of carbon in body-centred cubic *α* iron, containing a maximum of 0·04% carbon at 695°C. It is soft, ductile and readily cold-worked.

Cementite

A hard brittle compound of iron and carbon with the formula Fe_3C. This may exist in the free state usually as a grain boundary film, or as a constituent of the eutectoid pearlite.

Pearlite

This is the eutectoid structure consisting of alternate laminations of ferrite and cementite. It contains 0·83% carbon and is formed by the breakdown of the austenite solid solution at 695°C.

Austenite

A solid solution of carbon in face-centred cubic γ iron, containing a maximum of 1·7% carbon at 1 130°C. It is soft and non-magnetic, and only exists in plain carbon steels above the upper critical range. It may, however, occur at room temperatures in certain alloy steels.

MECHANICAL PROPERTIES OF PLAIN CARBON STEELS

The mechanical properties of slowly cooled plain carbon steels will depend upon the proportion of each of the microconstituents present. The mechanical properties vary linearly from 0% carbon (100% ferrite) to 0·83% carbon (100% pearlite). After 0·83% carbon, free cementite appears in the microstructure and the linear relationship exists no longer.

Using the values stated in Table 4.2 it is a simple matter to construct a graph such as that shown in Fig. 4.6. It is not practicable to carry out a tensile test on a specimen of cementite and hence the results given for this constituent are only estimated values.

Microconstituent	*TS* N/mm^2	*El*	*HB*
Ferrite	340	40	100
Cementite	45	nil	650
Pearlite	930	5	280

Table 4.2. The Mechanical Properties of the Constituents Present in Slowly Cooled Plain Carbon Steels

It should be emphasised that the properties obtained in normalising (which involves air cooling) vary according to the thickness of section. Fully annealed steels (furnace-cooled) would give a softer and more ductile steel. A further limitation of the use Fig. 4.6 is that it does not take into account the effect of variation of other elements present in plain carbon steels. With these limitations in mind, a simple empirical formula can be obtained from the slope of the TS curve in Fig. 4.6. This gives the relation between carbon

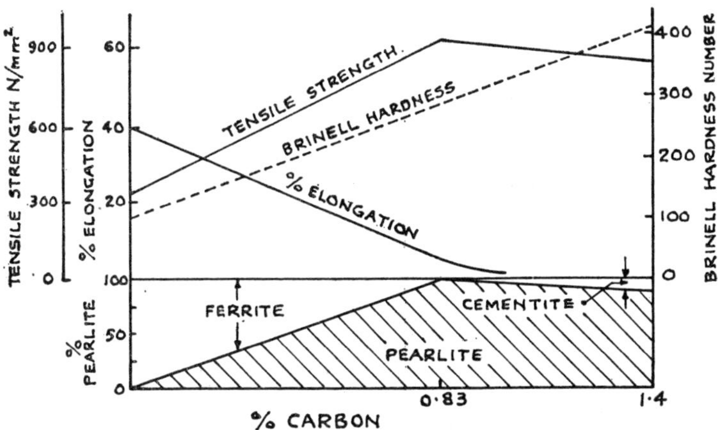

Fig. 4.6. The Effect of Carbon on the Microstructure and Mechanica Properties of Plain Carbon Steels in the Normalised Condition

content and tensile strength for a hypo-eutectoid normalised steel as,

$$y = 700x + 340$$
$$y = \text{TS in N/mm}^2$$
$$x = \text{carbon}\%$$

An empirical formula for estimating mechanical properties is not reliable, but provided its limitations are realised it provides a useful guide to the engineer.

RELATIONSHIP BETWEEN CARBON CONTENT AND MICROSTRUCTURE

It is evident that microscopic examination of slowly cooled plain carbon steels can be used to estimate the approximate carbon content. The following formulae can be readily derived for hypo-eutectoid steels.

$$\%\text{Pearlite} = \frac{\%\text{ Carbon} \times 100}{0{\cdot}83}$$

$$\%\text{Ferrite} = \frac{(0{\cdot}83 - \%\text{Carbon}) \times 100}{0{\cdot}83}$$

EFFECT OF OTHER ELEMENTS PRESENT IN PLAIN CARBON STEELS

In addition to carbon, all plain carbon steels contain the following elements:

Manganese —up to 1%
Silicon — „ „ 0·3%
Sulphur — „ „ 0·05%
Phosphorus— „ „ 0·05%

The separate effects of each of these elements are:

Manganese

This is an essential constituent since it ensures freedom from blow holes and combines with the sulphur present (see note on effect of sulphur below). In general manganese raises the yield point, TS and impact test values. It increases the depth of hardening but also increases the tendency to distort or crack upon quench-hardening. For this reason the manganese content should be kept below 0·5% in medium and high-carbon steels, which have to be heat-treated in this manner.

Silicon

In most commercial mild steels the silicon content is of the order of 0·1–0·2%. In these amounts it has little direct effect on the mechanical properties. In high-carbon steels it should not exceed 0·2% since it assists the breakdown of cementite into ferrite and graphite.

Sulphur

Sulphur in steel may exist in two forms:

(a) As manganese sulphide inclusions

These are soft dove-grey inclusions which are readily elongated in the direction of working (Fig. 4.7(*a*)).

(b) As ferrous sulphide inclusions

These occur as a brown grain boundary film (Fig. 4.7(*b*)). It is hard and brittle and possesses a low melting point thereby giving rise to cracking during hot- and cold-working of the steel. In order to avoid the formation of ferrous sulphide inclusions a manganese: sulphur ratio of at least 5:1 is maintained in plain carbon steels.

In general the sulphur content should be kept below 0·05% but certain free-cutting steels contain about 0·2% of sulphur and

1·5% manganese. This ensures the formation of numerous manganese sulphide inclusions which aid machinability.

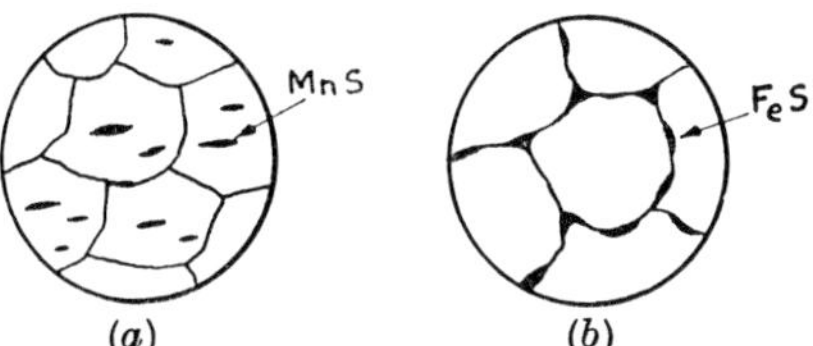

Fig. 4.7. (a) Manganese Sulphide Inclusions
(b) Ferrous Sulphide Inclusions at Grain Boundary

Phosphorus

Phosphorus has a pronounced tendency to segregate in steel. Hence the average composition should be kept below 0·05% to prevent the appearance of the brittle compound Fe_3P as a separate constituent.

CLASSIFICATION AND APPLICATIONS OF PLAIN CARBON STEELS (Table 4.3)

Carbon%	*Type*	*Use*
0·07–0·15	Dead Mild Steel	Hot and cold-rolled strip for pressings. Rod and wire for nails, rivets and mattresses. Solid-drawn tubes.
0·15–0·30	Mild Steel	Case-hardening steels. Boiler and ship's plate; steel sections, e.g. joists, channels, angles.
0·3–0·6	Medium-Carbon Steel	Forgings for general engineering purposes. Connecting rods. Axles. Crankshafts. Fish-plates.
0·6–0·8	High-Carbon Steel	Railway rails and tyres. Laminated springs. Wire ropes. Cast-steel die blocks. Band saws. Small forging dies.
0·8–1·4	Carbon Tool Steel	0·85–0·95% C. Small cold chisels. Shear blades. Punches. 0·95–1·1% C. Drills. Axes. Files. Hand saws. 1·1–1·4% C. Razors. Turning and planing tools. Drills.

THE COMPLETE IRON-CARBON DIAGRAM

The so-called complete iron-carbon diagram extends only to 6·68% carbon, which is the carbon content of cementite. The diagram is, therefore, more strictly the iron/iron carbide diagram. We have already considered the usefulness of the diagram in the study of steels. The diagram is also useful in the study of cast irons, which contain approximately 2·5–3·75% carbon. The structure and properties of cast irons will be discussed in Chapter 9.

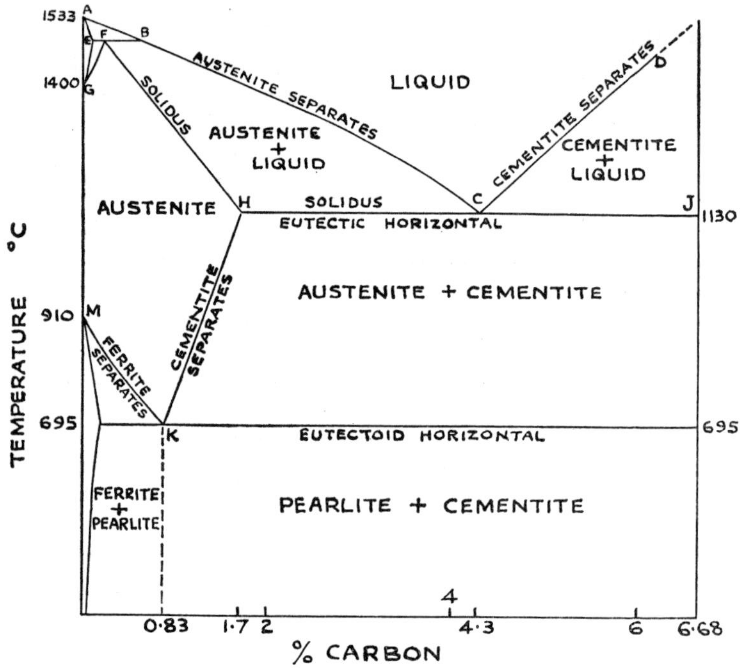

Fig. 4.8. The Complete Iron-carbon Diagram

The complete iron-carbon diagram may appear complicated at first sight, but may be considered as being made up of several simple basic diagrams, similar to those considered in Chapter 3. The main portions are as follows:

1. The Peritectic Portion (Fig. 4.9)

This occurs at 1,492°C. when δ solid solution containing 0·05%

carbon reacts with liquid containing 0·55% carbon to form a new solid solution, austenite, containing 0·18% carbon.

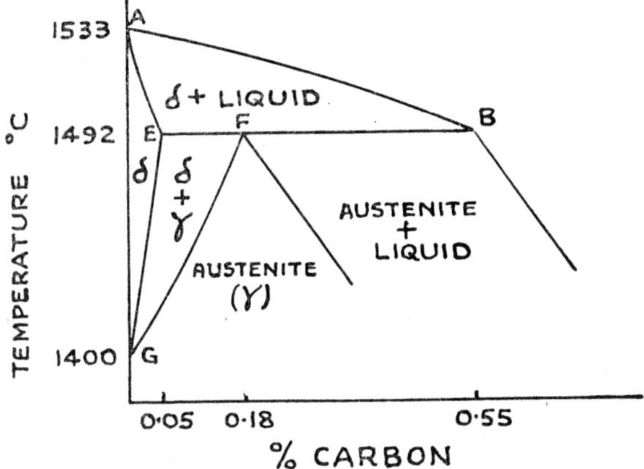

Fig. 4.9. The Peritectic Portion of the Iron-carbon Diagram (enlarged)

2. The Eutectic Portion

Liquid containing 4·3% carbon solidifies at 1,130°C. to form a eutectic consisting of austenite + cemenite.

3. The Solid-solution Portions

Two solid solutions are formed as previously mentioned on page 34. Austenite is a solid solution of carbon in γ iron, containing a maximum of 1·7% carbon at 1 130°C. Ferrite is a solid solution of carbon in α iron, containing a maximum of 0·04% carbon at 695°C.

4. The Eutectoid Portion

Austenite containing 0·83% carbon transforms at 695°C. to give a eutectoid consisting of ferrite and cementite, known as pearlite.

The lines MK and HK denote the beginning of the precipitation of ferrite and cementite respectively from austenite. The liquidus curves comprise AB, BC, and CD and the solidus curves AE, EF, FH, HC, and CJ.

5. Heat-treatment of Plain Carbon Steels

The mechanical properties of plain carbon steels can be varied considerably by heat-treatment. This is due to the structural changes which occur during the heating and cooling of such steels.

There are four conventional methods of heat-treatment:

1. Annealing
2. Normalising
3. Hardening
4. Tempering

In addition there are processes such as Martempering and Austempering depending upon the constant temperature (isothermal) transformations of austenite. The heat-treatment ranges for plain carbon steels are shown in Fig. 5.1.

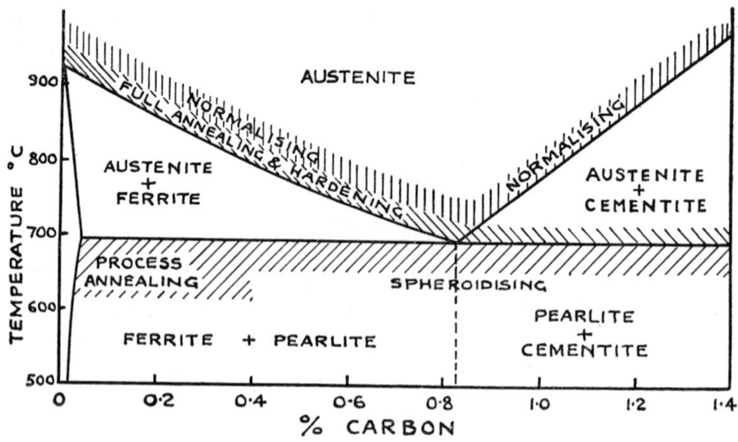

Fig. 5.1. Heat-treatment Ranges for Plain Carbon Steels

1. ANNEALING

Annealing is a general term applied to several softening operations, e.g. (*a*) process annealing, (*b*) full annealing, and (*c*) spheroidising.

1. Process Annealing

Process or sub-critical annealing is carried out on cold-worked low-carbon steel sheet or wire in order to relieve internal stress and to soften the material. The steel is heated to 550°–650°C., which is just below the lower critical point (Fig. 5.1).

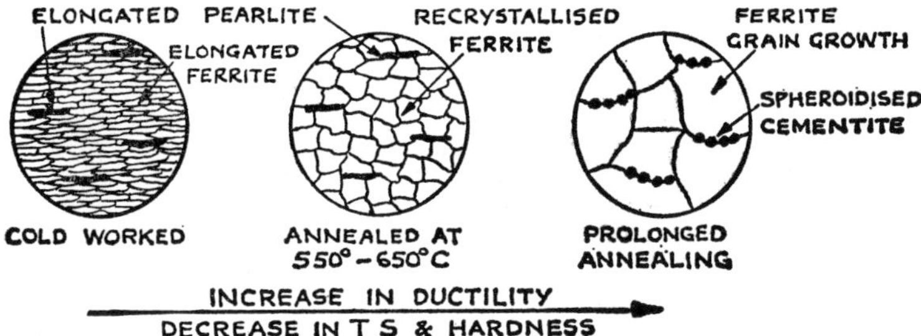

Fig. 5.2. Changes occurring during Process Annealing of Cold-worked Low carbon Steel

The changes taking place during process annealing are represented in Fig. 5.2. This is an example of recrystallisation annealing mentioned in Chapter 2. Prolonged annealing causes the cementite in the pearlite to 'ball up' or spheroidise. Ferrite grain growth also occurs and the annealing temperature and time should be closely controlled.

2. Full Annealing

Full annealing is carried out on hot-worked and cast steels in order to obtain grain refinement in combination with high ductility. Compared with normalising, it produces a softer steel with better machinability.

For hypo-eutectoid steels the treatment involves heating the steel to 30°–50°C. above the upper critical point, holding it at this temperature for a time depending on thickness, followed by slow cooling, usually in the furnace. For hyper-eutectoid steels the temperature is about 50°C. above the lower critical point.

If the temperature of annealing is not closely controlled, certain defects may occur, e.g. overheating, burning or under-annealing.

Overheating

If the steel is heated to above the correct annealing temperature, or if it is maintained too long at the annealing temperature, austenite grain-growth will occur. Upon cooling from this temperature, ferrite is deposited first at the grain boundaries and then along certain crystallographic planes giving rise to a structure such as that shown in Fig. 5.3(*a*). This type of structure is often known as a Widmanstatten structure, since it was first observed by Widmanstatten in meteorites. The structure is associated with weakness and brittleness, but can be remedied by reannealing to the correct temperature.

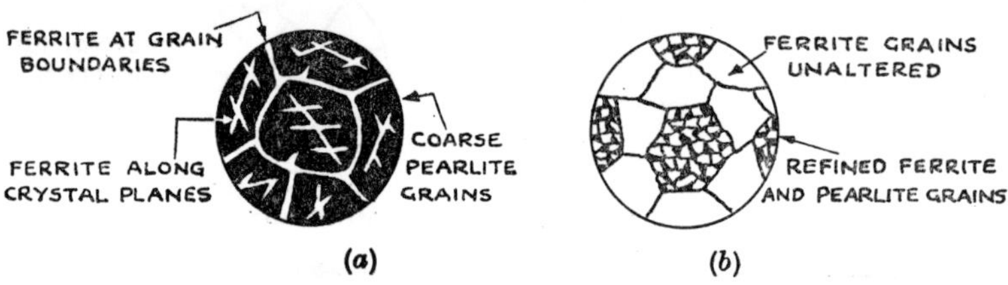

Fig. 5.3. Typical Microstructure of (*a*) Overheated Steel
(*b*) Under-annealed Steel

Burning

If a steel is heated far above the upper critical point to temperatures approaching the solidus, fusion and subsequent oxidation occur at the grain boundaries. Brittle films of oxide are formed which make the steel unsuitable for further use. The steel is said to be 'burnt' and must be remelted.

Under-annealing

Under-annealed structures are more frequently observed in the heat-affected zones of the parent metal in the welding of mild steel. The steel is heated to within the critical range, that is, between the upper and lower critical points. At this temperature the original pearlite will have changed to several small austenite grains. Upon cooling, ferrite is deposited at the austenite grain boundaries, the residual austenite transforming to pearlite at the eutectoid temperature. The resulting structure is similar to that shown in Fig. 5.3(*b*).

3. Spheroidising Annealing

High-carbon steels may be softened by annealing at 650°–700°C. (just below the lower critical point, Fig. 5.1), when the cementite of the pearlite balls up or spheroidises. The resulting structure is one of cementite globules in a ferrite matrix. In this condition the steels can be cold drawn and possess good machinability. Spheroidisation is more readily carried out on a fine pearlite structure when fine globules of cementite are obtained. Large globules, although producing a softer structure, present difficulties in machining and produce a poor surface.

2. NORMALISING

For hypo-eutectoid steels, normalising consists of heating the steel to 30°–50°C. above the upper critical point, holding it at this temperature for a time depending upon the section thickness, followed by cooling in still air.

Normalising produces maximum grain refinement, and consequently the steel is slightly harder and stronger than a fully annealed steel. However, the properties obtained in normalising will vary with section thickness. Large sections, which cool very slowly, may exhibit properties very similar to those of a fully annealed steel.

It will be noted from Fig. 5.1 that the normalising temperatures for hyper-eutectoid steels are above the upper critical points since a sufficiently fast air cooling prevents grain boundary precipitation of the cementite.

3. HARDENING

Hardening of hypo-eutectoid steels involves heating to 30°–50°C. above the upper critical point, holding the steel at this temperature for a time depending upon the thickness, followed by quenching in water, brine or oil.

The effect of cooling rate on the transformation temperature and products of austenite is shown in Fig. 5.4.

With very slow cooling the austenite transforms to lamellar pearlite. Increasing the cooling rate depresses the transformation temperature, giving a finer, harder pearlite, until a second transformation occurs at 150° to 350°C. when martensite is formed. When a certain cooling rate, known as the *critical cooling rate,* has been exceeded, the austenite transforms direct to martensite. Martensite is the hardest structure in a given steel, and therefore to harden a steel fully the critical cooling rate must be exceeded.

The critical cooling rate is lowered with increasing carbon and alloy content. With certain alloy steels the critical cooling rate may be sufficiently low to enable full hardening to be obtained by oil quenching (oil-hardening steels) or even air cooling (air-hardening steels)

There is no sharp line of demarcation between the structures obtained with various cooling rates, but the following terms may be used.

Lamellar Pearlite

A coarsely laminated structure, consisting of alternate laminations of ferrite and cementite, obtained upon very slow cooling.

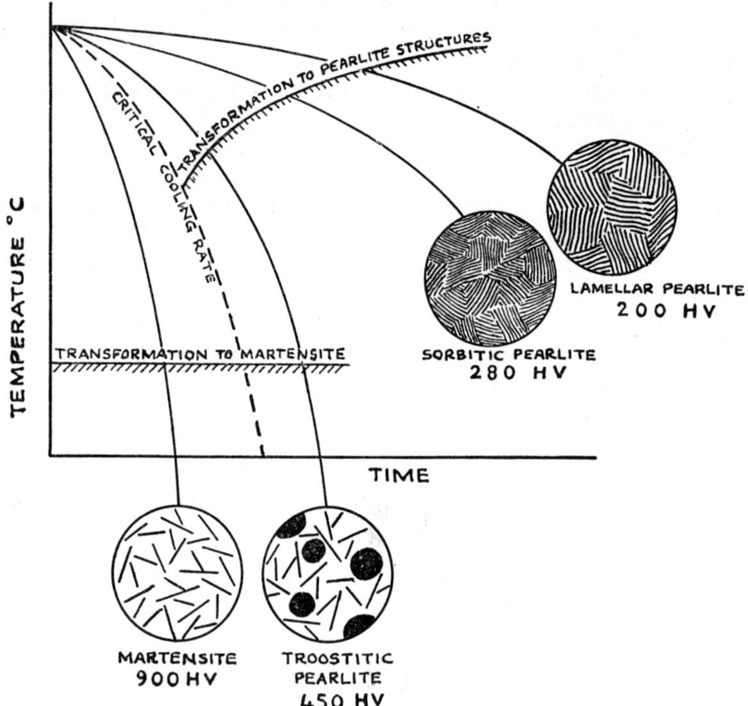

Fig. 5.4. Effect of Cooling Rate on the Transformation of a Eutectoid Steel

Sorbitic Pearlite

Sorbitic or *Fine Pearlite* is obtained by a faster rate of cooling. The laminations are closer together and may only be resolved with difficulty, if at all, at normal magnifications. Hardness and strength increase as the fineness of the pearlite increases.

Troostitic Pearlite

This consists of dark-etching nodules of pearlite, obtained by a more rapid rate of cooling such as oil quenching. At high magnifications the dark nodules are found to consist of laminations radiating from a central nucleus. The nodules are usually associated with some martensite. This structure may be known as *Primary Troostite* to distinguish it from the troostite obtained upon tempering, which we shall call *Secondary Troostite.*

Martensite

This is the hardest structure in a given steel, the hardness depending on the carbon content. It is seen as a needle-like structure under the microscope.

Fig. 4.2 shows that above the critical range steels have a face-centred cubic lattice, whilst at room temperatures they have a body-centred cubic lattice. Carbon is soluble in the face-centred cubic lattice forming the interstitial solid solution austenite. However, at room temperature the maximum solubility of carbon in body-centred cubic α iron is only 0·006%. Upon slow cooling the carbon in excess of this amount has time to separate out as cementite, which forms part of the pearlite structure. However, rapid cooling as in water quenching, prevents the separation of the cementite and a supersaturated solid solution of carbon in α iron, which is known as *Martensite*, is obtained. Martensite has a distorted lattice structure which will not permit the passage of dislocations, with the result that it is hard and non-ductile. The higher the carbon content, the greater the lattice distortion and the harder the martensite (Fig. 5.5). This structure is softened by reheating or tempering when the excess carbon is precipitated (Fig. 5.11).

The effect of the carbon content on the average hardening temperature and the critical cooling rate is shown in Table 5.1.

Carbon %	0·2	0·3	0·4	0·5	0·6	0·7	0·8	1·0
Average Hardening Temperature °C	870	850	830	810	790	770	760	760
Critical Cooling Rate °C./sec	1 200	1 000	800	700	600	550	500	400

Table 5.1. Effect of Carbon Content on the Average Hardening Temperature and the Critical Cooling Rate of Plain Carbon Steels

Hardenability. 'Mass Effect'

The terms hardness and hardenability should not be confused. Hardenability can be defined as the ability of a steel to be hardened by quenching, and is related to the depth and distribution of hardness throughout a section. It is not related to maximum hardness, which depends almost entirely on carbon content (Fig. 5.5), and not to any extent on alloy content, e.g. a water-quenched 0·9% carbon steel will possess a higher maximum hardness than a 3% nickel steel containing only 0·3% carbon. However, the latter will possess the greater hardenability since it will harden fully throughout a larger section.

For full hardening the actual cooling rate throughout the section must exceed the critical cooling rate for that steel. Actual cooling rate depends upon (*a*) the quenching medium, and (*b*) the diameter

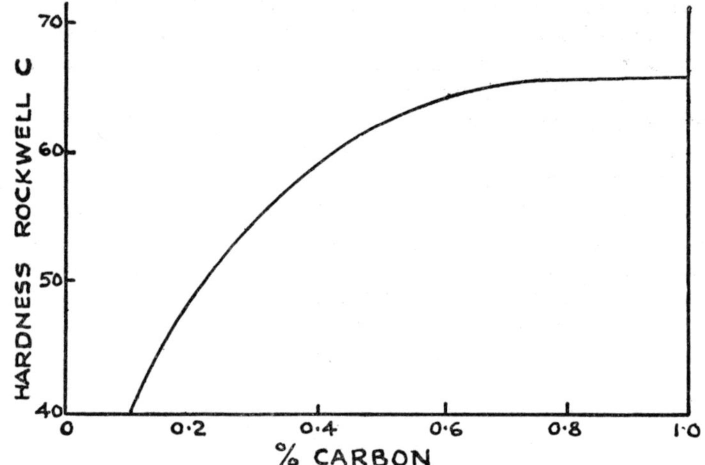

Fig. 5.5. The Effect of Carbon Content on the Maximum Hardness of Quenched Carbon Steels

of the bar or thickness of section. Plain carbon steels possess a high critical cooling rate and therefore large sections cannot be fully hardened throughout (Fig. 5.6).

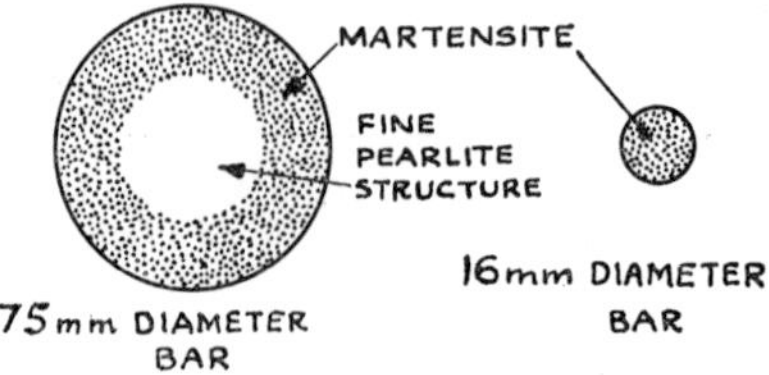

Fig. 5.6. The Effect of Diameter of the Bar on the Hardening of Water-quenched Plain Carbon Steels

It is obvious from Fig. 5.6 that the mechanical properties obtained by heat-treating a 75 mm-diameter bar will be different from those obtained in a 16 mm.-diameter bar of the same steel. In view of this, the term 'ruling section' has been introduced. If the 'ruling section' is exceeded, the stated mechanical properties will no longer apply, since hardening of the core will be incomplete. The critical cooling rate is reduced by the addition of alloying elements, and alloy steels are therefore not so liable to 'mass-effect' (Fig. 5.7).

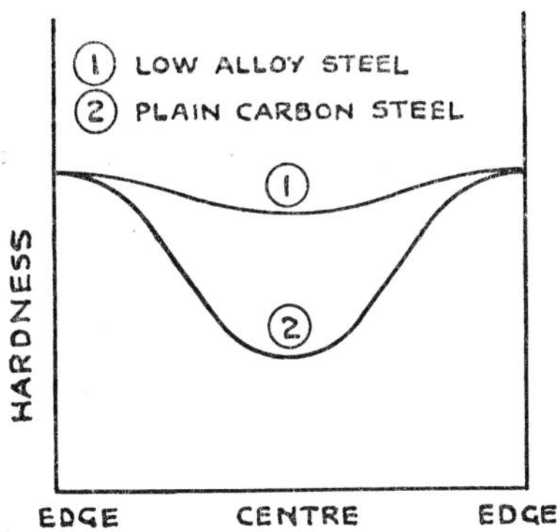

Fig. 5.7. Cross-section Hardenability of two Steels

The End-Quench Test for Hardenability

The Jominy end-quench test involves heating a test specimen of the dimensions shown in Fig. 5.8, to just above the upper critical range until it is fully austenitic. It is then transferred to a frame and quenched under standard conditions by a water jet which impinges on one end of the specimen only (Fig. 5.8).

The specimen cools very rapidly at the quenched end and progressively less rapidly towards the opposite end. When cold, a flat is ground along the side of the test piece about 0·4 mm deep, and hardness measurements are taken at intervals from the quenched end. The hardness values are plotted against the distance from the quenched end to give hardenability curves as shown in Fig. 5.9.

The curves may be used to compare the hardenability of steels by determining the distance to some selected hardness level or to the point of inflection of the curve. A comparison can also be made by examination of the microstructure along the end-quenched specimen. The point at which the steel develops a 50% martensite or the distance over which it develops a fully martensitic structure may be used.

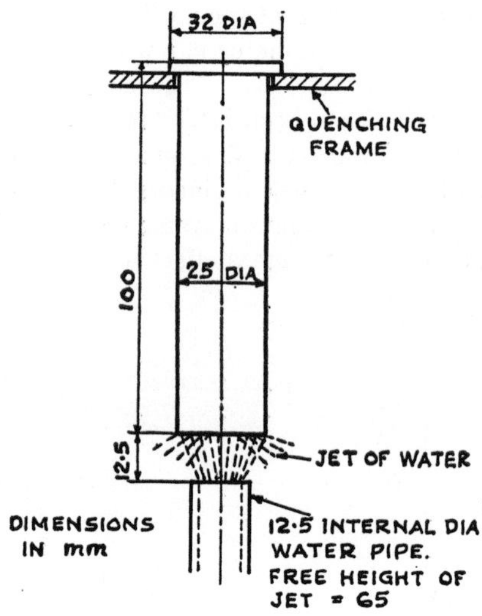

Fig. 5.8. Details of Jominy End-Quench Test

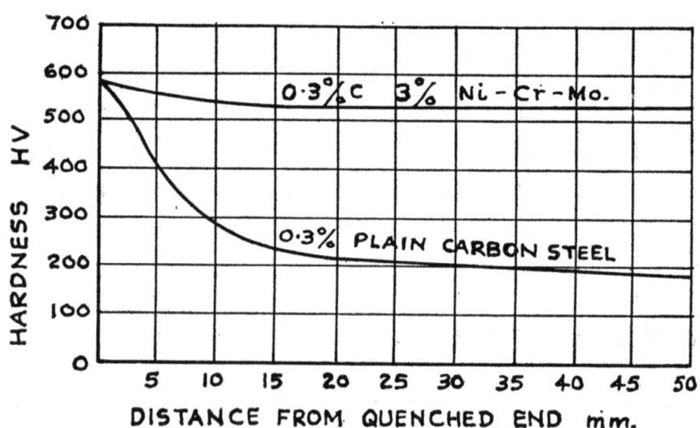

Fig. 5.9. Typical End-Quench Curves for Plain Carbon and Low Alloy Steels

4. TEMPERING

Tempering of hardened steel involves reheating to just below the lower critical point. The rate of cooling after tempering is not important for plain carbon steels. The objects of tempering are:

1. To relieve internal stresses induced by quenching.
2. To toughen the steel.

Tempering at 100°–200°C. is sufficient to relieve quenching stresses. When the temperature is in the region 200°–450°C. the martensite decomposes into ferrite and the precipitation of fine particles of carbide occurs. The fine granular structure formed is known as *secondary troostite* and must not be confused with the *primary troostite* of quenching which is laminated. This results in some toughening at the expense of hardness.

At higher temperatures 450°–650°C. the carbide particles coalesce thus producing fewer and larger particles which provide fewer obstacles to dislocations. The result is a decrease in strength and hardness while further increasing the toughness. This granular structure of carbide particles in a ferrite matrix is known as *sorbite* (Fig. 5.10) and is associated with maximum toughness.

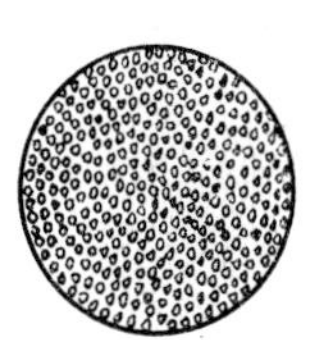

Fig. 5.10. Sorbite

Sorbite is the ideal microstructure for heat-treated components subject to dynamic stresses, e.g. axles, crankshafts and connecting rods.

It should be noted that there is no strict line of demarcation between *secondary troostite* and *sorbite*; the distinction is a matter of size of the precipitated carbide particles. These structures may also be known as *Tempered Martensite.*

The effect of tempering on the hardness of a quenched high carbon steel is shown in Fig. 5.11.

The Isothermal Transformation of Austenite

The behaviour of steels, when cooled at different rates from the austenitic state, may be understood more fully by studying the constant temperature or isothermal transformations of austenite. These transformations may be studied by heating several small specimens of the steel to above the upper critical range, followed by quenching rapidly into a constant temperature bath of molten lead or molten salt. The specimens are withdrawn at definite intervals and quenched in water (Fig. 5.12). This treatment converts any untransformed austenite to martensite and the amount of martensite

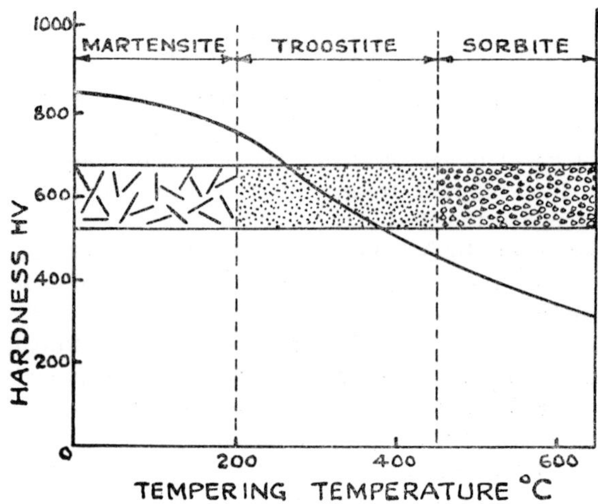

Fig. 5.11. Effect of Tempering on the Hardness and Microstructure of a Quenched Eutectoid Steel

is estimated by microscopic examination. In this way the extent of transformation at this temperature can be studied at various time intervals.

If the times for the beginning and end of transformation are plotted for each temperature, a curve is obtained (Fig. 5.13) known as the S curve, or T.T.T. (Time-Temperature-Transformation) curve for that steel.

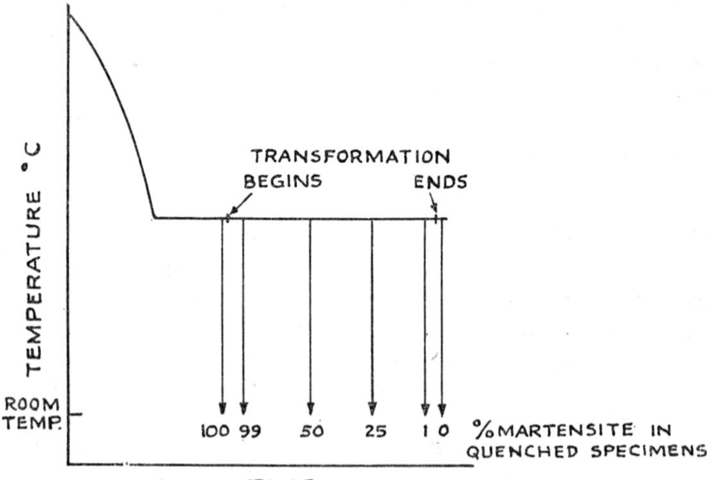

Fig. 5.12. Technique of Studying Isothermal Transformations

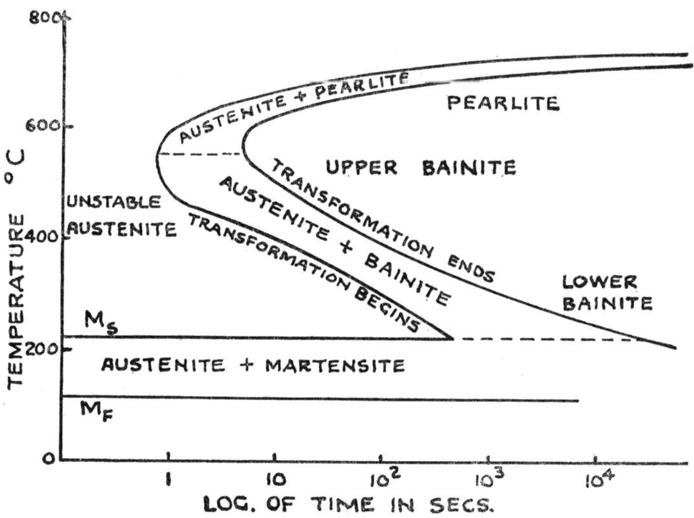

Fig. 5.13. Typical S-curve for Plain Carbon Steel

At higher temperatures the austenite will transform to pearlite, ranging from lamellar to troostitic pearlite as the temperature of transformation decreases. At intermediate temperatures a black rapid etching structure known as bainite is obtained, which at higher temperatures is feathery (*upper bainite*) and at lower temperatures needle-like (*lower bainite*). Lower bainite is not easy to distinguish from martensite, but is not as hard as the latter. The Ms and Mf temperatures denote the temperatures for the beginning and end of the transformation to martensite. These temperatures will depend upon the composition of the steel. Below the Ms temperature the austenite will transform to white slow etching martensite.

Heat-Treatments Based on the 'S' Curve

The change from austenite (γ iron) to martensite (α iron) is accompanied by volume change and since the cooling rate will vary across the section thickness, distortion and quenching cracks are likely to occur upon direct hardening (Fig. 5.14(*a*)). The processes of austempering and martempering are an attempt to avoid the distortion which is usually associated with the martensite reaction.

Martempering

The steel is heated to above the upper critical point and quenched into a salt bath maintained at a temperature just above the Ms temperature (Fig. 5.14(*b*)). It is held in the bath until the temperature is uniform throughout the section, when it is cooled in air. Martensite is thus formed with a minimum of distortion and danger of cracking. The time required for equalisation of temperature throughout the section should not exceed that for the beginning of the transformation to bainite. It is therefore limited to relatively small-section work, particularly of complicated design.

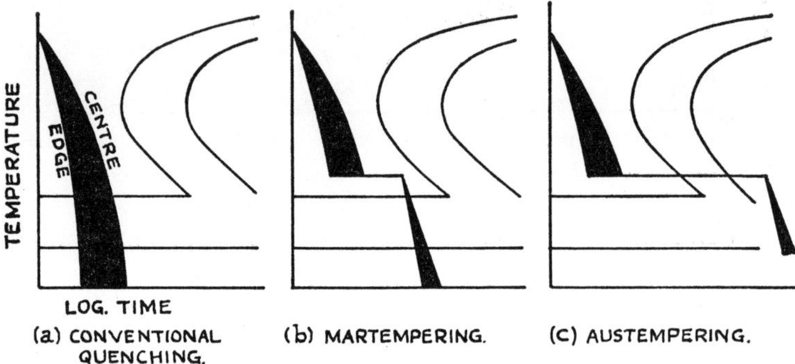

Fig. 5.14. Heat-treatments based on the S-curve.
Cooling Curves shown for Edge and Centre of Steel being treated

Austempering

The steel is heated to above the upper critical point and quenched into a molten salt or lead bath, maintained at a temperature in the bainitic region (250°–500°C.). It is held in the bath until the austenite has transformed to bainite, when it is cooled to room temperature at any desired rate (Fig. 5.14(*c*)).

The tensile and impact properties of austempered low- and medium-carbon steels are generally inferior to those of fully hardened and tempered steels. However, austempering is an advantage when dealing with work of complicated section, which might distort or crack when quenched in the conventional manner.

PRACTICAL ASPECTS OF HEAT-TREATMENT

Furnace Atmospheres

The control of furnace atmosphere is particularly important in the heat-treatment of steel in order to avoid such defects as oxidation or scaling, decarburisation (resulting in a soft skin) or carburisation (resulting in a hard skin).

In gas- and oil-fired furnaces, where the products of combustion come into contact with the steel being heated, it is essential to control the extent of the combustion of the fuel. Depending upon the volume of air used to burn the fuel, furnace atmospheres may be divided into three main types, namely oxidising, reducing, or neutral. An oxidising atmosphere contains an excess of oxygen and may give rise to scaling and decarburisation of the surface of the steel. A reducing atmosphere contains an excess of carbon monoxide, due to insufficient air being used for combustion. Although scaling is avoided, this may result in carburisation of the surface. A neutral atmosphere contains no free oxygen and little, if any, carbon monoxide. Such an atmosphere, which is not harmful, is the result of complete combustion, which is difficult to achieve in practice.

In gas-fired muffle furnaces (Fig. 5.15) the products of combustion do not make contact with the steel being treated. In these

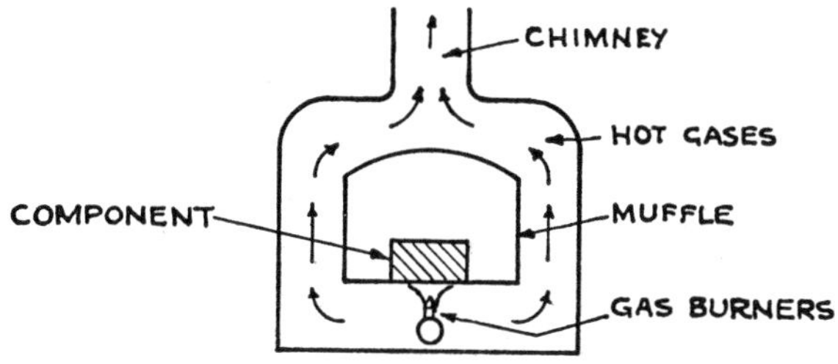

Fig. 5.15. Section through Gas-fired Muffle Furnace

furnaces, as in electric furnaces, the steel is in contact with an air atmosphere. Scaling and decarburisation will therefore take place, particularly at the higher temperatures. In such cases, artificial atmospheres may be used in heat-treatment furnaces. The air is purged out of the furnace and replaced by a controlled artificial atmosphere, at just above atmospheric pressure to prevent air entering the furnace. Various controlled atmospheres are in use, such as partially burnt town gas or partially burnt anhydrous 'cracked' ammonia.

Liquid Baths

Freedom from scaling and uniformity of heating can be obtained by immersing the steel in molten salt or molten metal baths. Various salt analyses are available for use at the temperatures involved in all heat-treatment operations. Neutral salts are used for such heating operations, but for liquid cyaniding (Chapter 6) carburising salts are employed.

Quenching Media

The chief quenching media, in order of their cooling rates, are brine solutions, water, oil and air. Water quenching is essential for the full hardening of plain carbon steels. However, upon quenching, an insulating blanket of steam tends to form between the metal and the water, thus appreciably reducing the cooling rate. The steel should therefore be agitated in the bath after quenching. As the temperature of the water in the quenching tank rises, the quench is less effective, since the tendency to form steam around the component increases. Brine solutions (e.g. sodium chloride in water) give a more drastic quench for two reasons. First, the solution has a higher boiling point, and secondly, the salt is effective in removing the scale, thus facilitating contact between clean steel and the quenching medium. However, such drastic quenching is not always necessary and may result in distortion or cracking. Alloy steels, which have a lower critical cooling rate, may be effectively hardened by oil quenching or even air cooling.

Quenching Cracks and Distortion

In addition to the normal thermal contraction, certain volume changes occur rapidly and unevenly throughout the section, when steel is quenched from the austenitic region. The change from austenite (face-centred cubic lattice) to martensite (body-centred cubic lattice) is accompanied by an expansion. This change occurs first at the outside of the section, which cools more quickly than the centre. In the centre of a large section, a contraction may occur where a troostitic pearlite structure has formed. Internal stresses are set up, which are likely to give rise to distortion or even cracking. The cracking may occur some time after quenching and occasionally during tempering.

Quenching Technique

Distortion can be minimised by bearing in mind a few simple points. For example, cylindrical specimens should be quenched vertically, flat sections edgeways, and the thicker section of a non-uniform component should enter the quenching bath first.

The Relationship of Design to Heat-treatment

Distortion and cracking in heat-treatment can frequently be traced to bad design of the component. Steel articles which have to be heat-treated should be designed so that the cooling rate is as uniform as possible throughout the section. This can be achieved by avoiding large variations in cross-section. Sharp changes in section should be avoided by providing generous fillets. Fig. 5.16(*a*) shows a badly designed punch. In the improved design (Fig. 5.16(*b*)) the thickness of

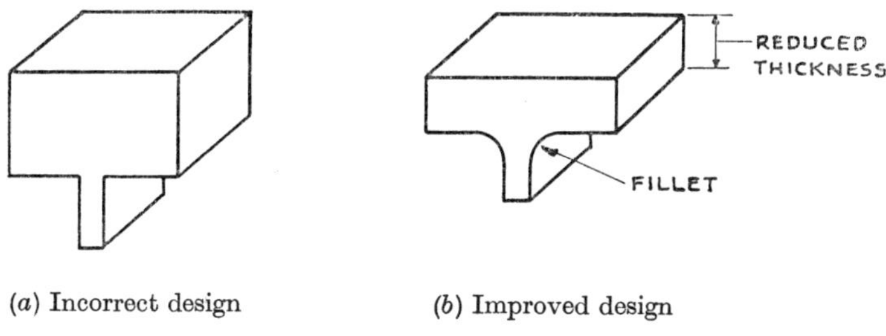

(*a*) Incorrect design (*b*) Improved design

Fig. 5.16. The Effect of Design in Heat-treatment

the horizontal portion has been reduced and a fillet provided, so as to give a more uniform cooling rate. In some cases holes may be drilled in the heaviest sections to reduce the mass without any adverse effect on the application of the article.

Temperature Measurement in Heat-treatment

Temperature measurement in the heat-treatment of steels is usually carried out using the thermo-electric pyrometer which is usually fitted through the furnace wall.

The principle of the thermo-electric pyrometer is that when two dissimilar wires are joined to form a closed circuit, and a temperature difference is maintained between the two junctions, a voltage is generated in the circuit. The voltage produced will depend upon (i) the metals used and (ii) the temperature difference between the hot and cold junctions. If the temperature of the cold junction is kept constant then the voltage produced is a measure of the temperature of the hot junction.

The thermo-electric pyrometer consists of three main parts as shown in Fig. 5.17.

1. **The Thermocouple**—This is the name given to the two dissimilar metals in wire form. The wires are insulated and are usually protected from contamination by a metal or silica sheath.
2. **Compensating Leads**—These connect the ends of the thermocouple to the temperature measuring instrument.

3. **The Temperature Measuring Instrument**—This is usually a millivoltmeter calibrated to read the temperature directly. The millivoltmeter usually possesses a high internal resistance so that any change in the resistance of the circuit due to temperature variation is negligible compared with the total resistance.

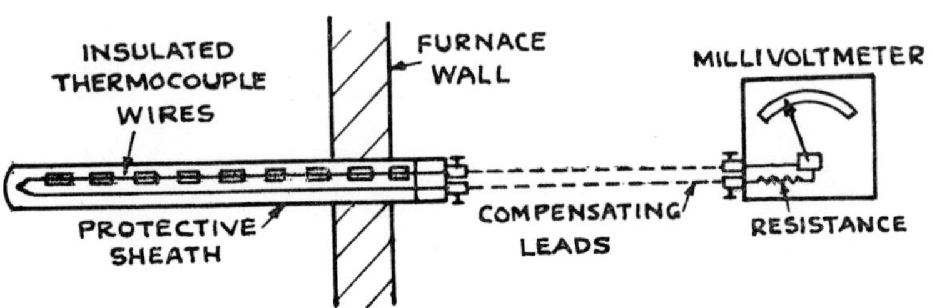

Fig. 5.17. Diagram Arrangement of Thermo-Electric Pyrometer as used in the Heat-Treatment of Steel

Thermocouple	*Maximum Temperature °C*		*Remarks*
	Continuous	*Intermittent*	
BASE METAL Copper-Constantan	400	500	5mV/100°C. approx
Iron-Constantan	900	1 100	5mV/100°C. approx Iron fails by oxidation
Chromel-Alumel	1 100	1 300	4mV/100°C. More suitable in oxidising atmospheres. Attacked by carbon, sulphur, and cyanide fumes.
RARE METAL Platinum—87% platinum/13% rhodium alloy	1 400	1 600	1mV/100°C. approx

Table 5.2. Details of Base-metal and Rare-metal Thermocouples

Most millivoltmeters used for this type of work are fitted with automatic cold junction compensation to allow for variations in the temperature of the cold junction.

The thermocouples used are of two types, namely (*a*) base-metal and (*b*) rare-metal couples.

Base-metal couples are cheaper and produce higher voltages for a given temperature rise. This enables the use of less sensitive temperature measuring instruments. They cannot be used to measure such high temperatures as the rare-metal couples and may in some cases be liable to oxidation or to attack by carbon, sulphur or cyanide fumes. Rare-metal couples are more expensive and require more sensitive measuring instruments. They are not readily oxidised and can be used at higher temperatures. Details of the more important thermocouples are given in Table 5.2.

Calibration of a thermocouple can be carried out by determining the millivoltage produced at a number of known temperatures, such as the freezing temperatures of pure metals and compounds.

Optical and radiation pyrometers may also be used. These are non-contact methods and may therefore be used to measure temperatures higher than those possible with a thermo-electric pyrometer, but are unsuitable for metals which are not 'red hot'. Their readings are affected by the presence of smoke and fumes and the character of the surface of the metal.

Maraging

The term 'maraging' is used to describe the age-hardening of certain high-alloy martensitic steels. The 18% NiCoMo maraging steel (page 71) is solution annealed by heating to 820°C. followed by air cooling to give a hardness of 290–320 HV. In this condition the alloy may be cold worked, if desired, since the carbon content is very low. Ageing at 480–500°C. for about 3 hours increases the hardness to 540–650 HV and gives a tensile strength of 1850–2150 N/mm^2. Hardening is due to a combination of ordering of the FeNiCo lattice structure and the formation of precipitates of the Ni_3Mo and Ni_3Ti type. In addition to high strength these steels are characterised by good notch-toughness, weldability and stress-corrosion resistance.

6. The Surface Hardening of Steel

It is frequently necessary for machine parts to possess a hard, wear-resisting surface and yet be sufficiently tough to withstand dynamic stresses. These properties are obtained by treating a tough steel, so as to increase the hardness of its surface. There are four important processes of surface hardening, namely:

1. Case hardening
2. Nitriding
3. Flame hardening
4. Induction hardening

Processes (1) and (2) involve a change in chemical composition of surface, whereas (3) and (4) involve only a change in microstructure by local heat-treatment.

CASE HARDENING

In case hardening, a low-carbon steel is heated to above the upper critical point, in a carbon-rich material. In this way the carbon content of the surface is increased to about 0·85% carbon. When followed by heat-treatment, a hard wear-resisting surface is obtained. The carbon-rich material may be solid, liquid or gaseous.

(a) Solid carburising (pack carburising)

The steel components (finish-machined or a little oversize) are packed in a solid carburising mixture in steel boxes, whose lids are sealed with clay, to exclude air. The box is heated at 900°–950°C. for three to eight hours, depending upon the depth of case required. After carburising, the components are allowed to cool slowly in the box. The solid carburising mixture may consist of charcoal or powdered coke, together with an energiser, such as barium carbonate. Various proprietary compounds are available on the market for this purpose. Any portion of the surface which is not required to be carburised may be copper-plated to a thickness of 0·075 mm.

The exact mechanism of carburisation is not fully understood, but the barium carbonate provides oxygen to oxidise part of the carbon to carbon monoxide. The carbon monoxide dissociates at the surface of the work, according to the chemical equation $2\ CO \rightarrow CO_2 + C$, and the freshly produced or 'nascent' carbon is absorbed by the steel. The rate of diffusion of the carbon into the steel will depend upon the carburising temperature, and the composition of the steel and of the carburising mixture.

In general, case depths of between 0·05 mm–1·55 mm are usually obtained by this method. When carburising at 900°C., about one hour at temperature should be allowed for each 0·25 mm of case depth.

(b) Liquid carburising

The steel components, contained in a wire basket, are immersed in a bath of molten salt, maintained at a temperature of 870°–950°C. The parts are usually quenched upon withdrawal. The salt mixture usually contains sodium cyanide (20–45%) together with sodium carbonate and sodium chloride.

Heating is rapid and uniform, and distortion is minimised. The process is economical for thin cases on small parts, and is used mainly for cases of up to 0·25 mm on mild-steel components subjected to light loads. Such case depths can be obtained in about forty-five minutes. The maximum economical case depth is about 0·75 mm, which may be obtained in about three hours.

A certain amount of nitriding also occurs, in addition to carburising. The presence of sodium chloride 'activates' the cyanide, giving rise to an increase in carburising action and a decrease in the nitriding action.

It is important to emphasise the dangers involved in the use of cyanide for hardening. Salt baths should be fitted with a hood to remove all fumes. Care should be taken to avoid burns due to splashing, and to prevent cyanide from entering the blood stream via cuts or sores.

(c) Gas carburising

Carburising can be conveniently carried out by heating the component to approximately 900°C. in a suitable gaseous atmosphere. The gases usually employed include treated town gas, ethane, propane and butane. The case depths vary from 0·25–1·0 mm and carburising times are usually up to four hours.

HEAT-TREATMENT OF CARBURISED STEEL

If the carbon content of the case, after carburising, is more than approximately 0·85% carbon, some free cementite may be deposited at the grain boundaries, on cooling slowly from the carburising temperature. This would result in embrittlement and the risk of 'flaking' or 'peeling' of the case in service. Since carburising involves prolonged heating at a high temperature, grain growth will occur in both the core and the case, and heat-treatment is necessary in order to obtain optimum properties. The carbon content of the case (0·85%) is different from that of the core (0·2% carbon), consequently a double heat-treatment is required (Fig. 6.1).

(a) Core Refining

The steel is first heated to just above the upper critical point for the core (approx. 870°C.), when the core consists of fine-grained austenite. After soaking at this temperature it is then quenched in water or oil, to produce a core structure consisting of small particles of ferrite, embedded in martensite.

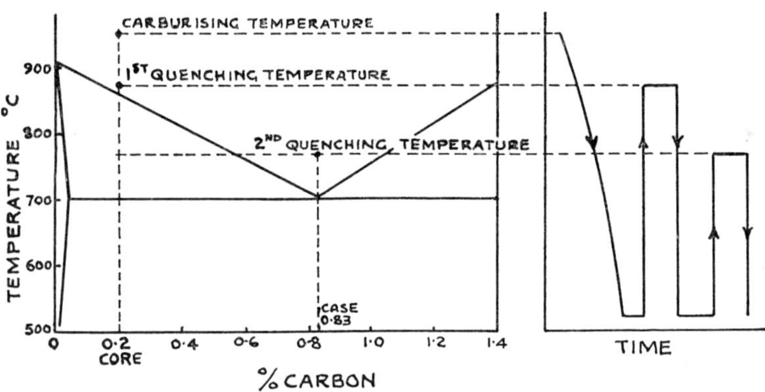

Fig. 6.1. Heat-treatment of Carburised Steel

However, 870°C. is well above the usual quenching temperature for a steel containing 0·85% carbon and consequently, after this treatment, the case will consist of a coarse martensite which is hard and brittle. A further treatment is therefore necessary to refine the case.

(b) Case Refining

The steel is heated to just above the critical point for the case

(760°C.) when the case consists of fine grains of austenite. Subsequent quenching from this temperature results in a fine-grained martensite which is hard, but not excessively brittle.

At 760°C. the core will consist of fine grains of austenite in a ferrite matrix. Upon quenching from this temperature, a structure of small martensite particles in a matrix of ferrite and troostitic pearlite is obtained. An improvement in core toughness can be obtained by rapid heating through the range 650°–760°C. followed by quenching without soaking. This treatment reduces the amount of martensite in the core.

A final tempering at 150°C. is advisable to relieve quenching stresses.

Case-hardening Steels

Mild steels and low alloy steels of the nickel or nickel-chromium type are chiefly employed for case-hardening.

Case-hardening Mild Steels

The carbon content may vary from 0·1% where maximum core toughness is required to 0·3% carbon, where higher load-carrying capacity is required. The manganese content should be approximately 0·6–0·9%. Manganese aids carburising and increases the depth of hardening, but the amount should be controlled, since it increases the tendency to distortion and cracking during heat-treatment.

Alloy Case-hardening Steels

These steels usually contain about 0·15% carbon and 2–5% nickel. Chromium may also be present such that the Ni:Cr ratio is approximately 3:1. The effect of nickel in case-hardening steels may be summarised as follows:

1. Nickel reduces the critical cooling rate so that oil quenching may be substituted for water quenching, with less risk of distortion and cracking.
2. Grain growth, during carburising, is restrained and in consequence the core refining treatment may, in certain cases, be eliminated.
3. The tendency towards cracking during grinding and flaking of the case is considerably reduced.
4. Improved yield point and tensile strength is obtained without any reduction in toughness.
5. The rate of diffusion of carbon is slightly reduced, consequently the time required for a given case depth is increased.

The effect of chromium is to increase the wear resistance of the case. Improvement in strength is also obtained without serious loss of toughness. 'Mass-effect' is considerably reduced by the use of low alloy nickel-chrome steels.

NITRIDING

A hard wear-resisting surface can be produced by using nitrogen instead of carbon, as the hardening agent. This process, known as nitriding, involves heating the finish-machined and heat-treated steel at 500°C. for 2–4 days in a gas-tight container, through which ammonia gas is circulated. The ammonia partially dissociates into nitrogen and hydrogen at the surface of the work, and the 'nascent' nitrogen diffuses into the surface, forming hard nitrides.

Plain carbon steels are not suitable for nitriding, since iron nitrides are formed which make the case too brittle. Special alloy steels are employed for nitriding, e.g. those of the *Nitralloy* type which contain 0·2–0·5% carbon, 1½% chromium, 1% aluminium and 0·2% molybdenum. Hard nitrides of chromium and aluminium are formed in the surface layers. Chromium nitride diffuses to a greater depth than aluminium nitride, consequently hardness falls off more gradually from the surface to the core. The presence of molybdenum gives grain refinement and improves the toughness of the core.

Heat-treatment is carried out prior to nitriding, in order to obtain a tough core. The full sequence of operations for a Nitralloy steel, similar to that mentioned, would be as follows:

1. Oil hardening from 900°C.
2. Tempering at 650°C.
3. Rough machining
4. Stabilising anneal at 525°–550°C. for up to 5 hours to remove machining stresses
5. Finish machining
6. Nitriding at 500°C. to give a case hardness of 1 050–1 100 HV. The usual case depth varies between 0·25 mm and 0·90 mm.

Typical applications of nitriding are die blocks, link pins, spindles, brake-drums, moulds for plastics, crankshafts, and printing dies.

Improved core properties can be obtained by the use of aluminium free steels. This is achieved at the expense of surface hardness which is reduced to about 750–900 HV. Examples of such steel are the 3% Cr-Mo and 3% Cr-Mo-V type, the latter containing 0·7–1·2% molybdenum and 0·1–0·3% vanadium.

Austenitic and ferritic heat-resisting and stainless steels can be successfully nitrided, following a special surface preparation. The nitriding temperature is usually 550°C. High-speed tool steels may be nitrided after the conventional treatment to increase the surface hardness. Certain alloy cast irons containing aluminium and chromium may be nitrided to give a surface hardness of 900–950 HV. The nitrided cast iron has a good corrosion and wear resistance and is used for cylinder linings and piston rings.

Compared with other processes of surface hardening, nitriding possesses a number of advantages, i.e.:

1. Distortion and quenching cracks are avoided since heat-treatment is carried out prior to nitriding. After nitriding, the parts are stress free.

2. A very hard surface (1 050–1 100 HV) can be obtained. The maximum hardness is usually obtained 0·03 mm–0·08 mm below the surface.

3. The hardness of the case is retained up to 500°C. in service. Carburised cases would begin to soften above 200°C.

4. The nitrided surface possesses a good resistance to corrosion, and for maximum corrosion resistance the parts should be used as they come from the nitriding container.

5. Nitrided steels have a good resistance to fatigue.

Nitriding would be a relatively expensive process for the surface hardening of small numbers of components, due to the initial high cost of plant and the necessity of using special alloy steels. However, the process is relatively cheap when large numbers of components are to be treated.

Carbonitriding is strictly a modification of the gas-carburising process. It involves the simultaneous carburising and nitriding of a plain carbon steel, by heating in a controlled atmosphere composed of ammonia and hydrocarbon gases. The temperature employed ranges from 650°–950°C., depending upon the steel and the depth of case required, but the usual temperature is about 820°–840°C. The steel may be quenched upon removal from the furnace. Subsequent heat-treatment is not necessary and there is less distortion, since a lower operating temperature is employed. The process is particularly applicable for hard, shallow, wear-resisting cases for automobile components.

FLAME HARDENING (THE SHORTER PROCESS)

The steel surface is heated to just above the upper critical point, by means of a moving oxy-acetylene torch, followed immediately by quenching, using a water spray attached to the torch.

The steels used should contain 0·4–0·6% carbon, in order to

ensure hardness of the surface whilst retaining reasonable toughness of the core. With 0·45% carbon, a surface hardness of 600–650 HV is obtained, whilst the depth of hardening is usually about 3·0 mm–3·8 mm.

The process is applied to gears, spindles, cams, clutches, sprockets, pinions, and the worms for worm-reduction units.

Before hardening, the parts should be stress free, and a low-temperature stress-relief anneal is usual after hardening.

INDUCTION HARDENING (THE TOCCO PROCESS)

The surface to be hardened is surrounded by a perforated inductor block, through which a high-frequency current is passed. The surface is quickly heated by the induced eddy currents to just above the upper critical temperature in about 3–5 seconds. Quenching is then carried out by spraying with water through the holes in the inductor block. A case depth of approximately 3·2 mm is obtained by this method, using similar steels to those used for flame hardening.

If the process is closely controlled, due to the speed of operation grain growth, decarburisation and distortion are prevented.

7. Alloy Steels

The properties of plain carbon steels have already been discussed in Chapter 4 and it will be realised that such steels have many limitations:

1. It is not possible to obtain a tensile strength of greater than 700 N/mm^2, if reasonable toughness and ductility are also desired.
2. Plain carbon steels are liable to 'mass effect'. Large sections cannot therefore be effectively hardened.
3. Drastic water quenching is necessary for full hardening, with consequent risk of cracking and distortion.
4. Plain carbon steels have a poor resistance to corrosion and to oxidation at elevated temperatures.

In order to overcome these limitations and to meet the specific requirements of engineers, alloy steels have been developed. The principal alloying elements are nickel, chromium, manganese, molybdenum, silicon, tungsten, vanadium, cobalt and copper.

EFFECT OF ALLOYING ELEMENTS

The effect of an alloying element may be one or more of the following:

1. It may strengthen the steel by going into solid solution. The formation of a solid solution (Chapter 3) results in distortion of the crystal lattice structure, which increases the hardness and strength.
2. When the solid solubility limit has been exceeded, the alloying element may form hard carbides. These carbides are usually associated with the cementite. Examples of carbide-forming elements are manganese, chromium, tungsten, and vanadium. Complex carbides are present in high-speed tool steels.
3. Certain alloying elements may give rise to the formation of graphite, which considerably reduces the strength, ductility and shock-resistance of the steel. Silicon, nickel and aluminium are

graphitising elements and it is usual to counteract their effect by adding carbide-forming elements also.

4. It may refine or coarsen the grain. Nickel tends to refine the grain, whilst silicon and chromium are grain-coarsening elements.

5. It may lower the carbon content of the eutectoid structure, pearlite. Certain low alloy steels may therefore possess more pearlite in their microstructure than that expected from the carbon content.

6. It may *raise* or *lower* the critical points, thereby stabilising *ferrite* or *austenite* respectively.

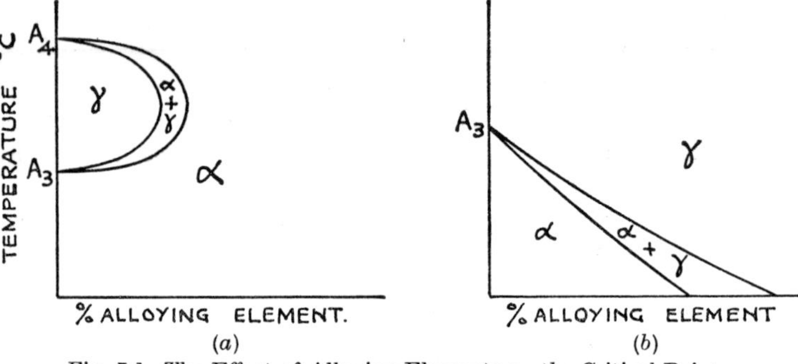

(*a*) (*b*)

Fig. 7.1. The Effect of Alloying Elements on the Critical Points
(*a*) A_3 Point raised
(*b*) A_3 Point lowered

In Fig. 7.1(*a*) the A_3 point is raised until it merges with the A_4 point to form what is known as a 'closed gamma loop'. Thus, above a certain percentage of alloying element, ferrite (α iron) is stable at all temperatures. Such alloys are not heat-treatable, since they have no critical points. Grain refinement can only be carried out by annealing after cold-work. Elements which stabilise ferrite are chromium, silicon, tungsten, vanadium and molybdenum.

In Fig. 7.1(*b*) the A_3 point is lowered and after a certain percentage of alloying element, austenite (γ iron) is stable at room temperatures. Elements which stabilise austenite are nickel, manganese, cobalt and copper. Such steels are austenitic on air cooling to room temperature and cannot be hardened by the normal quenching technique.

7. It may lower the critical cooling rate of the steel, thus increasing its hardenability. Alloy steels can be fully hardened in larger sections than plain carbon steels. In addition, oil quenching can be substituted for the more drastic water quenching, thus minimising the tendency to distortion and cracking. In certain alloy steels, the critical cooling rate may be reduced sufficiently to produce a martensitic structure on air cooling (air-hardening steels).

In addition to the general effects mentioned above, certain alloying elements may produce characteristic effects, e.g. the improvement in corrosion resistance by the addition of chromium. Tungsten and cobalt are noted for their effect on magnetic properties and vanadium for its effect in 'cleansing' a steel, making it free from inclusions.

CLASSIFICATION OF ALLOY STEELS

Alloy steels may be classified in many ways, but for the purposes of our study it is convenient to use a structural classification as follows:

(a) **Low Alloy Steels** which possess slowly cooled microstructures, similar to those of plain carbon steels in the same condition, namely pearlite, pearlite + ferrite or pearlite + cementite. These low alloy steels are often referred to as 'pearlitic' alloy steels.

(b) **High Alloy Steels** which possess slowly cooled microstructures, consisting either of martensite, austenite or ferrite + carbide particles. The reason for the presence of these structures has already been discussed.

In this chapter, an attempt will be made to study alloy steels from a structural standpoint. *The connection between the percentage of a single alloying element and the microstructure of the steel is affected by variations in cooling rate (size of section) and in the amount of other elements present, e.g. carbon. These limitations should be borne in mind in the following treatment of the subject.*

The effect of various alloying elements on the properties of steel will now be considered.

MANGANESE

Manganese goes into solid solution, increasing the strength and hardness, and also forms hard carbides. It lowers the critical cooling rate, thus increasing the hardenability of the steel and giving rise to air-hardening martensitic steels. Manganese lowers the critical points, thus stabilising the austenite. Manganese steels may be conveniently classified as follows:

Approximate Manganese %	*Type of Structure*
0–2	Pearlitic
2–12	Martensitic
12–100	Austenitic

Table 7.1. Structural Classification of Manganese Steels

This classification is only approximate as previously stated, but provides a useful basis for study.

Pearlitic Manganese Steels

These steels have to some extent replaced the more expensive pearlitic nickel steels. They contain about 1·5% manganese and 0·3–0·4% carbon, with in some cases 0·2–0·55% molybdenum. The presence of molybdenum reduces 'mass effect' and for larger section work the molybdenum content should be at the higher end of this range. The full heat-treatment involves oil hardening from about 840°C. followed by tempering at 600–650°C. In this condition they are used for shafts, gears, spindles, connecting rods and swivel arms. The manganese-molybdenum steels have mechanical properties roughly equivalent to those of a $3\frac{1}{2}$% nickel steel.

Martensitic Manganese Steel

These air-hardening steels have no commercial importance.

Austenitic Manganese Steel

Hadfield's manganese steel, containing 12–14% manganese and 1% carbon, was one of the first alloy steels to be discovered and produced commercially. It was developed in 1882 by Sir Robert Hadfield at Sheffield.

The steel is heat-treated by water quenching from 1 000°C., when the carbides are taken into solid solution to give a uniform solid solution of austenite (Fig. 7.2). In this condition the alloy is soft, as measured by the normal indentation tests. The Brinell Hardness Number is about 200. However, the alloy has excellent wear-resisting properties, since abrasion converts the surface layers into a hard structure with a Brinell Hardness Number of about 550. The steel is used for rock drills, crushers, railway points and dredging equipment. The alloy is difficult to machine but may be forged or hot-rolled.

Fig. 7.2. Microstructure of Hadfield's Manganese Steel. Water-quenched 1,000°C. Twinned Crystals of Austenite

NICKEL

Nickel has a marked strengthening effect on the steel, since it goes into solid solution and decreases the carbon content of the eutectoid. It is a graphitising element, but this effect may be counteracted by the presence of carbide-forming elements such as manganese or chromium. Nickel lowers the critical cooling rate, thereby increasing the hardenability of the steel. Oil- and air-hardening steels may be produced. It also lowers the critical range, thus stabilising austenite. Nickel steels may be conveniently classified as follows, bearing in mind the limitations previously mentioned:

Approximate Nickel %	*Structure*
0–8	Pearlitic
8–22	Martensitic
22–100	Austenitic

Table 7.2. Structural Classification of Nickel Steels

Pearlitic Nickel Steels

In practice such steels usually contain up to 5% nickel with 0·1–0·4% carbon. Those with a carbon content of 0·1–0·15% are suitable for case-hardening, whilst those containing 0·25–0·4% carbon are used in the heat-treated condition for parts subjected to dynamic stresses, e.g. crankshafts, axles and connecting rods.

The full heat-treatment usually involves oil hardening from 830°–860°C. followed by a tempering at 550°–650°C. to produce a tough steel. The minimum mechanical properties for typical nickel steels in the fully heat-treated condition are as follows:

Type	*YP* N/mm^2	*TS* N/mm^2	*El* %	*Izod joules*	*HB*	*Ruling Section mm*
3% Ni	550	770	17	54	223	63
	490	690	19	54	201	100
$3\frac{1}{2}$% Ni	630	850	15	54	248	63
	550	770	17	54	223	100

Table 7.3. Minimum Mechanical Properties of Typical Heat-treated Low Alloy Nickel Steels

Martensitic Nickel Steels

These include the new high-strength maraging steels (page 59), e.g. the steel containing 18% Ni, 9% Co, 5% Mo, 0·5% Ti, 0·03% max C.

Austenitic Nickel Steels

The austenitic nickel steels are not commercially important but there are numerous austenitic iron-nickel alloys (carbon-free) which find wide application. These iron-nickel alloys are used as low and controlled-expansion alloys and for their magnetic properties. Well-known industrial alloys of this type include *Invar, Kovar* and *Permalloy.*

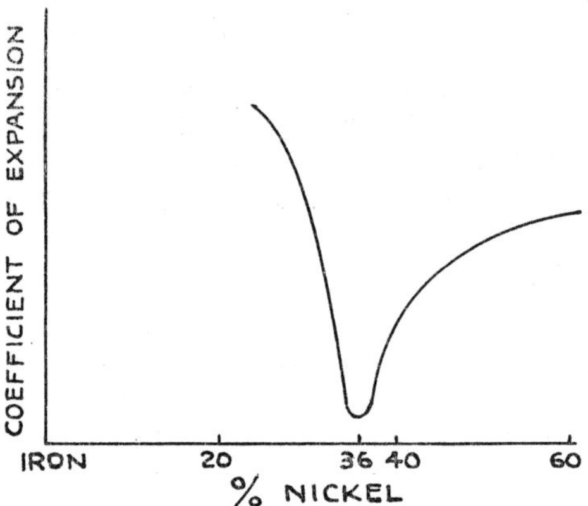

Fig. 7.3. Graph showing Relationship between Coefficient of Expansion and Nickel Content of Iron-nickel Alloys

Invar (36% Ni) has a very low coefficient of thermal expansion in the temperature range 0–100°C. (Fig. 7.3). It is used for watch and clock mechanisms, metal tapes, surveying instruments and thermostats.

Alloys with 30% nickel and up to 17% cobalt are used for making glass-to-metal seals in radio valves and other thermionic devices. These alloys have approximately the same coefficient of expansion as the harder glasses of the boro-silicate type and are known under such trade names as *Kovar, Nilo K,* and *Fernico.*

Permalloy (78·5% Ni) possesses a high magnetic permeability in weak magnetic fields (page 86).

CHROMIUM

Chromium goes into solid solution in the steel and also forms hard carbides. It lowers the carbon content of the eutectoid, and also the critical cooling rate, thereby increasing the hardenability of the steel. Chromium forms a 'closed gamma loop' and thus stabilises ferrite. It is a grain-coarsening element and therefore prolonged heating at high temperatures should be avoided. Chromium steels may be roughly classified as follows, bearing in mind the limitations previously mentioned:

Approx. Chromium%	*Structure*
0–2	Pearlitic
2–16	Martensitic
Above 16	Ferritic

Table 7.4. Structural Classification of Chromium Steels

Pearlitic Chromium Steels

These steels usually contain 1·0–1·5% chromium and 0·35–1·0% carbon.

The medium-carbon steels may be oil hardened and tempered at 600°–650°C. to give a tough structure suitable for parts subjected to dynamic stresses, e.g. gears, axles, crankshafts and connecting rods. Steels containing 1·5% Cr and 0·2% V are used for springs. Chrome-vanadium spring steels are tempered at 420°C. and have an ultimate tensile strength of 1 400–1 550 N/mm^2.

The high-carbon steels are used for ball and roller bearings, small cold rolls, shoes for ore crushers and other components requiring a hard surface. These steels contain a higher proportion of free cementite than that indicated by the carbon content, and after heat-treatment this will be present in the spheroidised form. One important alloy in this group is En 31 containing 1·0% carbon and 1·4% chromium which is used for ball bearings. This steel is oil hardened from 810°C. and tempered at 150°C. to give a Brinell Hardness Number of 850.

Martensitic Chromium Steels

Steels containing 1·0% carbon and 3·5% chromium are used in the oil-hardened condition for permanent magnets (page 85).

Complete resistance to atmospheric corrosion is obtained in steels with greater than about 12% chromium. This is due to the formation of a thin adherent oxide film on the surface which also provides resistance to corrosion by oxidising solutions. These 'martensitic' stainless steels will be discussed more fully in the next chapter.

Chromium also increases the resistance to scaling at elevated temperatures and steels such as *Silchrome* (8% Cr 3·5% Si) are used for automobile valves.

Ferritic Chromium Steels

These alloys containing 0·05–0·15% carbon and 16–30% chromium are usually known as the 'ferritic' stainless steels. They will be considered more fully in the following chapter.

NICKEL AND CHROMIUM

The alloying elements nickel and chromium are frequently used together and two main types of nickel chromium steel may be distinguished.

(a) Low Alloy Nickel-chrome Steels

These steels containing 1–5% nickel and 0·6–1·5% chromium are the most important alloys for general engineering purposes since a wide range of mechanical properties can be obtained by the appropriate heat-treatment.

(b) Austenitic Stainless Steels

These are usually based on the 18:8 Cr:Ni type and will be discussed in the following chapter.

LOW ALLOY NICKEL-CHROME STEELS

The reason for the wide use of nickel-chrome steels is that the beneficial properties given by each element are additive, whereas the disadvantages associated with the use of either element singly are counteracted by the presence of the other element. For example the strength, ductility and toughness associated with nickel steel are combined with the hardness and wear resistance of chromium steels. The grain coarsening tendency of chromium is counteracted by the grain refining tendency of nickel, whilst the graphitising tendency of nickel is counteracted by the carbide-forming tendency of chromium. These desirable properties are obtained by maintaining a Ni:Cr ratio of approximately 3:1.

The lower carbon steels (C = 0·10–0·15%) are suitable for case hardening, whilst those with 0·25–0·35% carbon are used in the heat-treated condition for parts subjected to dynamic stresses, e.g. crankshafts, axles, connecting rods, swivel arms, etc. The full heat-treatment involves oil hardening from 820°–850°C. followed by tempering at 550°–650°C.

Nickel-chrome steels are prone to a defect known as 'temper-brittleness'. This term usually refers to the drop in the Izod impact test value obtained by tempering in the range 250°–400°C., or by slow cooling through this range after tempering at 600°C. Such steels are not generally tempered in the range 250°–400°C. Temper-brittleness can be minimised by:

1. Oil quenching after tempering at 600°–650°C., or
2. The addition of 0·3–0·5% molybdenum to the steel. It is advisable to oil quench after tempering even if a Ni-Cr-Mo steel is used.

There is no complete explanation of temper-brittleness, which is only revealed by notched-bar impact tests.

Details of the minimum mechanical properties of typical nickel chrome steels in the fully heat-treated condition are given in Table 7.5.

Type	*YP* N/mm^2	*TS* N/mm^2	*El* %	*Izod joules*	*HB*	*Ruling Section mm*
3% Ni-Cr	800 590	1 000 770	14 17	47 54	293 223	63 150
3% Ni-Cr-Mo	900 680	1 080 850	12 16	40 54	311 248	100 150
3½% Ni-Cr-Mo	1 050 740	1 240 930	12 15	34 47	263 269	63 150
4¼% Ni-Cr-Mo	1 310	1 550	9	20	446	150

Table 7.5. Minimum Mechanical Properties of Typical Nickel-Chrome Steels in the Fully Heat-treated Condition

The 4¼% Ni-Cr-Mo steel is of the air-hardening type and is used for large-section work requiring freedom from distortion.

SILICON

Silicon lowers the carbon content of the eutectoid and forms a 'closed gamma loop'. It is a marked graphitising element and steels with more than 5% silicon are commercially useless due to the presence of graphite. It has only a small effect on the hardenability since the graphitising tendency exceeds the air-hardening tendency.

There are three important types of silicon steel, namely:

(a) **Silico-Manganese Steel** Si = 1·5–2·0% Mn = 0·6–1·0%
Used for manufacture of leaf springs.

(b) **Silicon-Steel** C = 0·07% Si = 4%
Possesses high magnetic permeability and electrical resistance and is used for transformer cores and the poles of dynamos and motors (page 85).

(c) **Valve Steels**
Silicon is a secondary alloying element in automobile valve steels such as *Silchrome* and *Valmax* which contain 8% chromium and 3·5% silicon (page 82).

TUNGSTEN

Tungsten is a strong carbide-forming element. These carbides do not readily go into solid solution even at high temperatures. Hardened tungsten steels resist tempering up to relatively high temperatures, hence the use of tungsten in high-speed tool steels. Tungsten refines the grain size and decreases the tendency to decarburisation during working. Tungsten increases coercive force and a steel containing 1% carbon and 6% tungsten is used for permanent magnets. Small amounts of tungsten are also present in certain heat- and corrosion-resisting steels.

MOLYBDENUM

Molybdenum is a strong carbide-forming element. It reduces 'mass-effect' and 'temper-brittleness' and inhibits grain growth. It is rarely used alone as an alloying element. Molybdenum improves the mechanical properties at high temperatures and 0·5% molybdenum is present in creep-resisting steels for use at high steam temperatures. It is used in high-speed tool steels, and heat- and corrosion-resisting steels.

VANADIUM

Vanadium is also a strong carbide-forming element and is used in high-speed tool steels. It is a grain-refining element and a strong deoxidiser. It ensures a clean steel by acting as a 'scavenger' for oxides and other inclusions.

8. Special Purpose Steels

Alloy steels have been developed to meet many requirements and in this chapter we shall consider some of the more important types of special alloy steels.

STAINLESS STEELS

Stainless steels were first discovered by Brearley in 1913 when he observed that a steel containing 13% chromium exhibited a good resistance to atmospheric corrosion. The corrosion resistance is due to the formation of a thin stable protective oxide film on the surface of the steel. This film also provides resistance against corrosive attack by solutions of an oxidising nature.

The 13% chromium steel was the basis of the original stainless steel cutlery, but since that time many varieties of stainless steel have been developed. Stainless steels may be classified into three main types:

1. Martensitic
2. Austenitic
3. Ferritic

Martensitic Stainless Steels

The martensitic stainless steels can be subdivided into three main types.

(a) Stainless Irons

$$C = 0{\cdot}07\text{–}0{\cdot}10\%$$
$$Cr = 13\%$$

These alloys are usually air cooled or oil quenched from 950°–1 000° C. followed by tempering at 650°–750°C. In the hardened and tempered condition the alloy can be cold-worked and readily machined. It is readily welded and is used for turbine blades, split pins and rivets. The lower carbon alloys are used for domestic articles such as forks and spoons.

Typical mechanical properties for stainless iron in the hardened and tempered condition are:

YP *N/mm²*	*TS* *N/mm²*	*El*	*HB*
230–390	460–620	25–35	140–180

(b) Stainless Steels

C = 0·2–0·4%
Cr = 13%

Hardening is carried out by oil quenching from 950°C. In this condition the steel possesses maximum corrosion resistance since all the carbides are in solid solution. Tempering at 500°–750°C. toughens the steel but lowers the corrosion resistance due to the precipitation of chromium-rich carbides. These alloys are difficult to weld and their corrosion resistance is decreased by cold-working. A high polish is necessary for maximum corrosion resistance.

The applications of these steels depend upon the carbon content.

0·2% carbon—used for valves and piston rods but should not be used in contact with copper-base alloys or graphitic packings owing to electrolytic action resulting in corrosion.

0·3% carbon—used for table cutlery, surgical and other instruments requiring a sharp cutting edge. The steel is usually tempered in the range 150°–180°C. for these applications.

0·4%–1·0% carbon—used for springs and ball bearings which work in a corrosive environment.

(c) High Chromium—Low Nickel Stainless Steels

e.g. S.80 containing 0·1% carbon, 18% chromium, 2% nickel. This steel combines good corrosion resistance with good mechanical properties obtained by heat-treatment. The heat-treatment involves oil hardening from 950°C. followed by tempering at 650°C. and in this condition typical mechanical properties are:

YP *N/mm²*	*TS* *N/mm²*	*El*	*HB*
620–770	850–1 000	12–22	250–320

It may be used in contact with graphite packings and copper-base alloys and typical applications include pump rods and shafts, regulator valves, blow-off cocks, turbine blading and aircraft fittings.

Austenitic Stainless Steels

There are numerous austenitic stainless steels but most are based on the 18:8 Cr:Ni type. These alloys cannot be hardened by quenching. Quenching from 1 050°C results in the carbides being taken into solid solution to give a fully austenitic structure (Fig. 8.1). In this state the alloy is soft and non-magnetic, and exhibits maximum corrosion resistance. Typical mechanical properties for a steel containing 0·1% carbon, 18% chromium and 8% nickel after water quenching are:

Fig. 8.1. Microstructure of 18:8 Cr:Ni Stainless Steel, Water-quenched 1,050°C. Twinned Crystals of Austenite

YP N/mm^2	*TS* N/mm^2	*El*	*RA*	*HB*
250–300	610–770	40–60	40-60	160–200

The chief uses are for chemical plant construction, breweries, dairies, food processing and for domestic and decorative purposes.

If austenitic stainless steels are heated in the temperature range 500°–800°C. precipitation of chromium-rich carbides occurs at the grain boundaries resulting in a decrease in the corrosion resistance. This is due partly to the resulting chromium deficiency in the grains and partly to the electrolytic action between the carbide particles and the matrix.

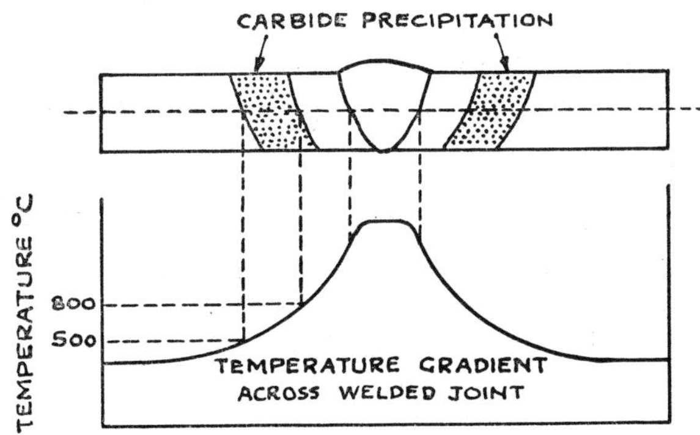

Fig. 8.2. Carbide Precipitation in the Heat-affected Zones of the Plate in the Welding of 18:8 Stainless Steel

During welding the heat-affected zones of the plate are maintained in this temperature range for sufficient time to allow carbide precipitation to occur. The phenomenon is therefore usually referred to as 'weld-decay'.

'Weld-decay' may be minimised or prevented by:

1. Water quenching from 1,050°C. after welding. This treatment takes the carbides back into solid solution but is not always practicable, particularly for large welded vessels.

2. Reducing the carbon content. This reduces the amount of carbide precipitated at the grain boundaries. 'Extra low carbon' stainless steels are now available with a maximum of 0·03% carbon.

3. The addition of 'stabilising' elements such as titanium and niobium. These elements have a stronger affinity for carbon than that possessed by chromium. The precipitated carbides will therefore consist chiefly of titanium carbide or niobium carbide and consequently chromium impoverishment of the grains does not occur. Such steels are known as 'stabilised' stainless steels and should always be specified for welded construction. The amount of stabilising element is given by the following formulae:

$$\% \text{ titanium} = 6 \times (\% \text{ carbon in excess of } 0{\cdot}02\%)$$
$$\% \text{ niobium} = 10 \times (\% \text{ carbon in excess of } 0{\cdot}02\%)$$

4. The addition of ferrite-forming elements, e.g. silicon, molybdenum, tungsten. These form ferrite 'islands' in the microstructure in which the carbide is precipitated when the steel is heated in the range 500°–800°C. In this way the formation of a continuous brittle film is avoided.

Steels of the 18:8 type may also contain molybdenum (3%) and copper (2%) to give resistance to corrosion by various solutions. When molybdenum is added for this purpose the nickel content may be raised to 10% in order to counteract the tendency to form ferrite. Machinability may be increased by the addition of 0·2% selenium. An austenitic stainless steel for deep drawing purposes contains approximately 12% each of nickel and chromium.

Development of Sigma Phase

Steels of the 18:8 type which, due to various additions, have an austenite-ferrite structure are often referred to as duplex steels. If a duplex steel is heated in the range 500°–900°C. for some time a brittle constituent known as SIGMA PHASE is formed from the ferrite. When duplex steels are required for use at elevated temperatures the amount of ferrite should therefore be kept at a minimum.

Ferritic Stainless Steels

The ferritic steels containing 0·05–0·15% carbon and 16–30%

chromium are not amenable to heat-treatment. The structure can only be refined by recrystallisation after cold-working. The microstructure consists of ferrite grains and carbide particles. These steels are used for parts where mechanical properties are secondary to corrosion resistance and good workability.

HEAT-RESISTING STEELS

Heat-resisting steels are required for a wide variety of applications such as superheater tubes and pipes in steam power plant, aero-engine and automobile valves, furnace conveyors, retorts, oil cracking units, gas turbines, and glass-making machinery. Heat-resisting steels should possess the following properties:

1. Good creep resistance

Creep is the slow plastic deformation which occurs under prolonged loading to elevated temperatures (page 176).

2. Resistance to oxidation and scaling

The presence of chromium, silicon and aluminium produces hard adherent films on the surface which protect the alloy against further attack. The alloy should also be resistant to chemical attack by vapours and gases which might exist in service.

3. Structural stability

From the dimensional standpoint it is essential that components for use at high temperatures should have a stable structure. Carbide precipitation at the grain boundaries and the formation of sigma phase should be avoided. Spheroidization of carbides leads to a marked reduction in creep resistance. Since absolute dimensional accuracy is not possible some allowances have to be made in design.

4. Specific properties

Properties relating to a particular application must also be considered, e.g. machinability, weldability, fatigue properties, and coefficient of thermal expansion.

In order to meet these requirements a number of heat-resisting steels have been developed. These steels may be classified as follows:

1. Low Alloy steels (0·5% Molybdenum)
2. Chromium-silicon Valve steels
3. Plain Chromium steels (12–30% Chromium)
4. Austenitic Chromium-Nickel steels

1. Low Alloy Steels

The main applications for these steels are for superheater tubes and pipes in steam-power plants where service temperatures are in the range 400°–550°C. Typical compositions of suitable steels are those containing 0·5% Mo, 0·5% Mo + 1% Cr, 0·5% Mo + 0·25% V, 0·5% Mo + 2·25% Cr. These additions are all carbide-forming elements and such alloys undergo precipitation hardening when tempered in the range 600°–700°C. after normalising.

2. Valve Steels

Chromium-silicon steels such as *Silchrome* (0·4% C, 8% Cr, 3·5% Si) and *Valmax* (0·5% C, 8% Cr, 3·5% Si, 0·5% Mo) are used for automobile valves. They possess a good resistance to scaling at a dull red heat, although their strength at elevated temperatures is relatively low. The use of motor fuel treated with tetraethyl lead has led to the development of *Silchrome XB* steel (En 59) with a better corrosion resistance. This is a Cr:Si:Ni steel containing approximately 20% Cr, 2% Si, 1·5% Ni. For aero-engines and marine diesel engines the 13/13/3 nickel-chromium-tungsten valve steel (En 54) is usually employed.

3. Plain Chromium Steels

These include the martensitic chromium steels (12–13% Cr) and the ferritic chromium steels (18–30% Cr).

Plain chromium steels are noted more for oxidation resistance at high temperatures than for their strength which is not high under such conditions. The maximum operating temperature for the martensitic steels is about 750°C., whereas for the ferritic steels it is about 1 000°–1 150°C. Such steels have a good resistance to sulphurous atmospheres.

4. Austenitic Chromium-Nickel Steels

The austenitic steels combine good mechanical properties at high temperatures with good scaling resistance. These alloys contain a minimum of 18% Cr and 8% Ni stabilised with titanium or niobium. Other carbide-forming elements such as molybdenum and tungsten may also be added. Carbide-forming elements improve creep strength by giving rise to precipitation hardening alloys. Such alloys are suitable for use up to 1 100°C. Austenitic steels are suitable for gas turbine discs and blades.

HIGH-SPEED TOOL STEELS

Plain carbon tool steels are unsuitable for high-speed machining since the heat produced by frictional effects would temper the hard martensitic structure. To meet the needs of modern machining practice special alloy high-speed tool steels have been developed.

The best known high-speed tool steel is that containing 0·6% carbon, 18% tungsten, 4% chromium, and 1% vanadium, known as the 18:4:1 type. These elements form hard carbides which resist tempering, thus improving the hardness at red heat.

Steels of the 18:4:1 type may also contain 5–12% cobalt. Cobalt improves 'red-hardness' and gives rise to a greater degree of 'secondary hardening' (below). Cobalt high-speed steels are particularly suitable for cutting hard, sandy or scaley material such as sand castings, cast iron and heat-treated steels.

During the 1939–45 war supplies of tungsten from the Far East were not readily available and 'substitute' steels were developed in which molybdenum replaced part of the tungsten. A typical steel is that of the '4:6' type containing 4% molybdenum, 6% tungsten, 4% chromium and 1% vanadium.

Hardening of High-speed Tool Steels

In order to dissolve most of the carbides prior to quenching, a very high temperature of the order 1 250°–1 320°C. is required. However, at this temperature grain growth, decarburisation and incipient fusion may occur and it is necessary to adopt a special heat-treatment technique. This is usually carried out in a two-chamber gas-fired furnace, preferably using a controlled atmosphere. The steel is preheated in the upper chamber to about 850°C. After 'soaking' at this temperature it is transferred to the lower chamber which is maintained at 1 250°–1 320°C. depending upon the composition of the steel. The time in the lower chamber is usually 1–3 minutes for small tools. It is then usually oil quenched, when the microstructure will consist of particles of carbides in a matrix of martensite and retained austenite.

Secondary Hardening

The quenched steel can be further hardened by tempering at 550°–570°C. (Fig. 8.3). This phenomenon, known as 'secondary hardening', is thought to be due to the transformation of retained austenite to martensite. The higher the alloy content of the steel the more it responds to secondary hardening. Tempering at lower temperatures 350°–400°C. would result in softening (Fig. 8.3).

Cemented Carbide Tools

Cemented carbide tools are made by powder metallurgy and consist of hard tungsten carbide with or without the carbides of tantalum,

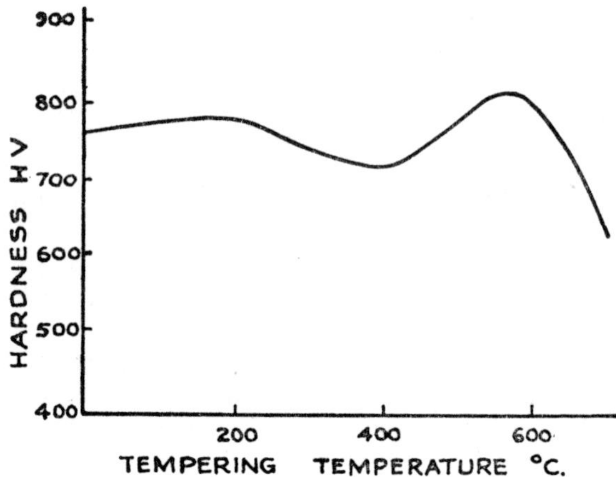

Fig. 8.3. Effect of Tempering Temperature on the Hardness of Quenched High-speed Tool Steel of the 18:4:1 Type

vanadium, molybdenum and titanium, bonded with 6–20% cobalt. The ingredients in powdered form are pressed to shape and sintered at 1 500°C. when bonding occurs aided by partial melting of the cobalt. Well-known examples are *Widia* and *Cutanit* which are capable of higher cutting speeds than the normal high-speed tool steels. They are used for machining cast iron, hard steel and non-metallic materials such as plastics. Due to the high cost, cemented carbide tools usually consist of a carbide tip or insert brazed on to a steel holder or shank.

Stellites

These are cobalt-base cast alloys containing elements such as chromium, tungsten and carbon. The structure is usually one of complex carbides in a cobalt and chromium matrix. They are used for tool tips, valves and valve seatings.

MAGNETIC ALLOYS

Magnetic alloys may be roughly divided into two classes according to whether or not they retain their magnetism. The former are referred to as magnetically 'hard' materials and are used for permanent magnets, whilst the latter are said to be magnetically 'soft' and are used for transformer cores, motor and generator armatures and other electrical equipment. A general connection exists between magnetic hardness and mechanical hardness. Treatments which increase mechanical hardness usually increase magnetic hardness.

Permanent-magnet Alloys

The earliest permanent-magnet alloy was the 1% carbon steel in the hardened condition. This was followed by the 6% tungsten steel and the 3–6% chromium steels. In 1920 Honda in Japan introduced a series of permanent-magnet alloys containing 3–35% cobalt with superior properties. These represent the limit of development for quenched martensitic steels.

In 1931 a new era in permanent-magnet alloys began with the development by Mishima in Japan of alloys containing nickel, aluminium and iron. These alloys, in common with the later alloys, undergo precipitation-hardening when tempered at 600°C. after air-cooling from 1,200°C. Development of such alloys, to which the name

Material	*Approximate composition*					*BH max. WbA/m³*
Martensitic steels						
Carbon steel	1% C					1 440
Chromium steel	1% C + 3·5% Cr					2 320
Tungsten steel	1% C + 6% W					2 400
3% Cobalt	1% C + 3% Co + 7% Cr					2 800
6% Cobalt	1% C + 6% Co + 7% Cr					3 520
15% Cobalt	1% C + 15% Co + 9% Cr					4 960
35% Cobalt	1% C + 35% Co + 5% Cr					7 600
Precipitation-hardening alloys	Al	Ni	Co	Cu	Ti	
Alni	13	24		3·5	Sometimes small additions	10 000
Alnico	10	18	12	6		13 600
Alcomax II	8	11	24	6		34 400
Ticonal	8	14	24	3		40 000
Alnico V	8	14	24	3		36 000
Columax	8	14	23	3		54 400

Table 8.1. Permanent Magnet Alloys

Alni (Table 8.1) was given, was continued in this country by the Permanent Magnet Association Ltd. The addition of cobalt and copper gave rise to the *Alnico* series of alloys which were developed in

Sheffield. Alloys such as *Alcomax*, *Ticonal* and *Alnico V* possess directional or anisotropic properties obtained by placing the alloy in a strong magnetic field during the hardening treatment. Alcomax magnets can be cooled during casting in such a way that their columnar crystals are developed in the same direction as the preferred axis of magnetisation. This type of alloy is known commercially as *Columax* and represents the most effective type of permanent magnet alloy available, based on the criterion of the value of the product BH maximum.

High Permeability Alloys

Magnetically soft materials should be readily demagnetised and retain as little magnetism as possible. Such alloys must have a high magnetic permeability, and must absorb a minimum of energy in an alternating magnetic field such as that experienced by a transformer core.

The earliest alloy used for this purpose was soft iron, followed by the iron-silicon alloys containing up to 4·5% silicon. Silicon iron is usually manufactured in the form of dead soft sheet for transformer core laminations. It is rolled and annealed so as to obtain grain growth which assists in obtaining the desired magnetic properties. Preferred orientation of the grains in the direction of rolling can be achieved by alternate cold rolling and annealing. This gives rise to superior magnetic properties in the sheet in the direction of rolling. Nickel-iron alloys such as *Permalloy* (78·5% Ni) and *Mumetal* (75% Ni) are noted for their high magnetic permeability in weak magnetic fields (Fig. 8.4). They are used as shields for submarine cables and for transformer cores.

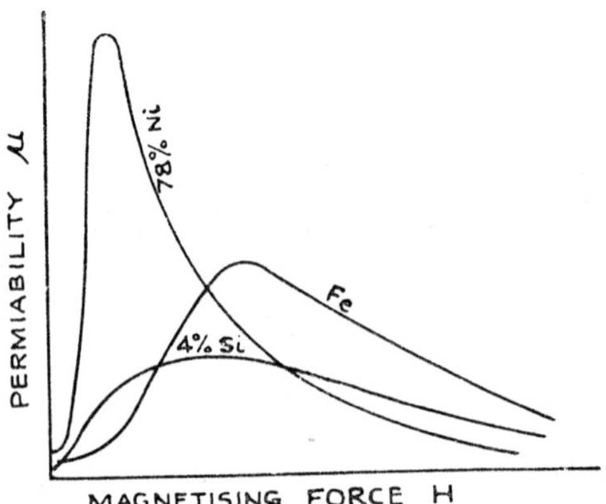

Fig. 8.4. Permeability Curves for 78% Nickel-iron Alloy, Silicon-iron and Soft Iron

9. Cast Iron

Cast iron consists basically of pig iron which has been remelted and cast, either alone, or blended with scrap iron or steel. Its melting point (approximately 1 150°–1 250°C.) is lower than that of steel and, in consequence, it can be more readily and cheaply melted, usually in a cupola furnace. The molten metal is very fluid and has the ability to take good casting impressions.

Like steel it always contains the five elements carbon, silicon, manganese, sulphur and phosphorus, but in larger proportions than exist in steel. A typical range of composition for cast iron is C = 2·8–3·6%, Si = 1·0–3·0%, Mn = 0·4–1·0%, S = 0·1–0·35%, P = 0·05–1·0%.

WHITE AND GREY CAST IRONS

The carbon in cast iron may exist in two forms, namely:

1. Combined carbon or cementite, Fe_3C
2. Free carbon or graphite formed by the breakdown of the cementite as follows:

$$\underset{\text{(Cementite)}}{Fe_3C} \longrightarrow \underset{\text{(Ferrite)}}{Fe} + \underset{\text{(Graphite)}}{C}$$

If the carbon is combined as cementite, the iron will be hard, brittle and unmachinable. A white fracture is apparent when the iron is broken, hence the name *white cast iron.*

If the carbon is free, in the form of graphite, the iron will be relatively soft and machinable and will give a greyish fracture (*grey cast iron*).

If only about half of the cementite has broken down into graphite the iron is referred to as *mottled iron,* due to the mottled appearance of its fracture.

The factors affecting the form of the carbon in cast iron are:

1. The rate of cooling
2. The chemical composition
3. Subsequent heat treatment

1. The Rate of Cooling

A high cooling rate tends to stabilise the cementite, giving a hard white iron, whereas a slow cooling rate assists the formation of graphite, thus producing a grey iron. The cooling rate will depend upon the section thickness and the type of mould used.

In castings of varying cross-section the thinner sections may consist of white iron and will be harder than the thicker sections, which will consist of grey iron. The cooling rate would be slower in a sand mould than that obtained in a metal mould.

2. Chemical Composition

(a) Carbon lowers the melting point as indicated by the iron-carbon diagram and increases the amount of graphite in the iron.

	White cast iron	*Grey cast iron*
The form of carbon	Combined as Cementite	Free carbon or Graphite
Chemical composition	Low silicon, high sulphur	High silicon, low sulphur
Cooling Rate in mould	Fast	Slow
Properties	Hard, brittle, unmachinable. HB = 400–500	Relatively soft and machinable. HB = 180–240
Typical uses	Ploughshares, chilled rolls, balls, stamp shoes, dies and wearing plates. Manufacture of malleable C.I.	Ingot moulds, automobile cylinders and pistons. Machine castings, water main pipes.

Table 9.1. White and Grey Cast Irons

(b) Silicon aids the formation of graphite and thus tends to produce a grey iron.

(c) Sulphur. The direct effect of sulphur is to stabilise the cementite, thus producing a white iron.

(d) Manganese combines with the sulphur to form manganese sulphide and thus indirectly aids the formation of graphite by its effect on the sulphur. However, the direct effect of manganese is to stabilise the cementite. This will occur only if the amount of Mn is greater than that required to combine with the sulphur (1 part S to 1·72 parts Mn).

(e) Phosphorus has no effect on the form of the carbon in cast iron. It does, however, improve the fluidity by the formation of a low-melting-point phosphide eutectic (Fe-Fe_3C-Fe_3P, melting point 960°C.).

3. Subsequent Heat-treatment

White-iron castings can be graphitised by an annealing treatment, as in the manufacture of malleable cast iron (dealt with later).

The distinction between grey and white cast irons may be summarised as in Table 9.1.

THE STRUCTURE OF GREY CAST IRONS

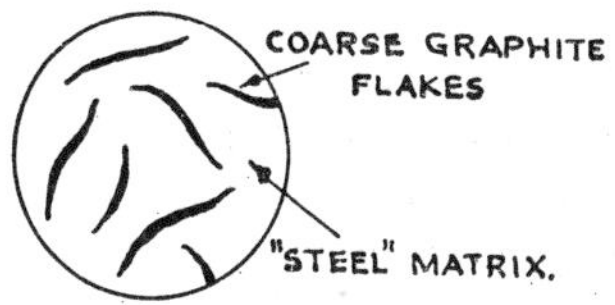

Fig. 9.1. Grey Cast Iron

Structurally, grey cast irons may be considered as being composed of a 'steel' matrix and coarse flakes of graphite (Fig. 9.1). The 'steel' matrix, usually pearlitic, has a tensile strength of approximately 770 N/mm², yet ordinary grey cast iron has a TS of only 180–230 N/mm², with very little shock resistance. This can be attributed to the presence of the graphite flakes, which have no strength, and act in the same way as cracks with sharp edges. This gives rise to points of stress concentration which decrease shock resistance and strength.

The strength of grey cast iron can, therefore, be increased in one of two ways, either by modifying the form of the graphite so as to eliminate its weakening effect, or by strengthening the 'steel' matrix. Using the first method, the presence of coarse flakes of graphite is avoided in the manufacture of malleable, inoculated and spheroidal graphite cast irons (Fig. 9.2), whilst the strength of the 'steel' matrix may best be increased by the addition of alloying elements.

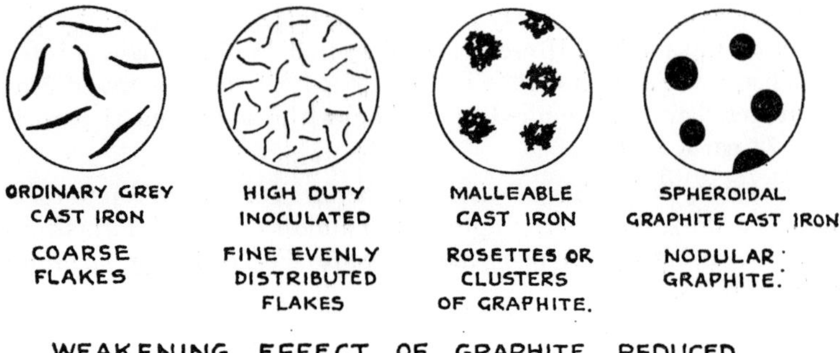

Fig. 9.2. Form of the Graphite in Various Grey Irons

MALLEABLE CAST IRON

Malleable cast iron is made by annealing white iron castings followed by controlled cooling. The total times involved in the annealing processes vary from 40 to 100 hours depending upon the composition, section thickness and the process employed. The two main types of malleable iron are Whiteheart and Blackheart (Table 9.2). The latter usually has a matrix structure of ferrite, but the process can be controlled to produce a matrix of pearlite. These pearlitic malleable irons can be heat-treated to produce desirable properties and can be surface hardened to 540 HV by flame and induction hardening.

The disadvantages of malleable cast iron are the length of time involved in the process, and the fact that the process is limited to small sections (usually less than 50 mm), since a white iron has to be produced initially.

Applications

Malleable iron is used for small structural components for automobiles such as hubs for lorries, differential housings, leaf-spring links, brake drums, wheel hubs, pedals and levers, and door hinges. It is also used for small parts in agricultural and textile machinery, gas and oil burners, pipe fittings, and plumbing supplies.

	Whiteheart	*Blackheart*	
Maximum annealing temp.	1 050°C.	950°C.	
Furnace atmosphere	Decarburising	Inert	
Object of annealing	Mainly decarburisation but some graphitisation	Graphitisation only	
Microstructure	 Rosettes of graphite in a ferrite-pearlite matrix. Thin sections will be completely decarburised giving ferrite only. In thicker sections the carbon content will increase from edge to centre	 Rosettes of graphite in a ferrite matrix. No variation in carbon content across the section. Process can be controlled to produce pearlitic matrix	
Mechanical properties		*Ferritic*	*Pearlitic*
TS N/mm^2	340–430	340–390	430–770
Elongation %	5–10	12–18	2–10
HB	120–240	110–120	170–270

Table 9.2. Malleable Cast Iron

INOCULATED HIGH DUTY CAST IRON

Castings with graphite in a fine evenly distributed flake form can be produced by the addition of a small amount of graphitising inoculant in the ladle just prior to casting. The inoculants used are calcium silicide (Meehanite process) and nickel shot and ferro-silicon (Ni-Tensyl process). The composition of the molten iron is such that it would normally produce a white iron. The effect of inoculation disappears upon remelting. The matrix structure consists entirely of fine pearlite. The approximate composition and mechanical properties of Meehanite and Ni-Tensyl inoculated irons are given in the following table:

Iron	*TC*	*Si*	*Mn*	*S*	*P*	*Ni*	*TS* N/mm^2	*HB*
Meehanite	2·5	1·3	0·8	0·14	0·1	—	310–390	260–300
Ni-Tensyl	2·8	1·5	0·8	0·6	0·1	1·5	340–460	280

SPHEROIDAL GRAPHITE CAST IRON

Castings with the graphite in nodular form can be obtained by adding a small amount of magnesium, in the form of nickel-magnesium alloy, to the ladle before casting. The alloy addition is of the order of 2%. The composition of the molten iron is such that it would normally cast white. There are two main types of S.G. iron, namely cast and annealed (Table 9.3).

S.G. cast iron with a pearlite matrix may be heat treated by quenching and tempering, induction or flame hardening, or by normalising. It is also possible to apply the magnesium treatment to alloy cast irons and so combine the beneficial effects of alloying with those obtained by having the graphite in the spheroidal form.

Applications

The present and possible applications of S.G. iron are numerous. Its use enables section thickness, and consequently weight, to be reduced. Toughness and ductility can be obtained in castings which are too thick for malleabilising. Such irons can replace steel castings and forgings. Applications include cast crankshafts, agricultural and marine castings, heavy machinery frames, hand tools, gas and water pipes.

ALLOY CAST IRONS

The chief elements alloyed with cast iron are nickel, chromium, copper and molybdenum. Of these, nickel is the most important and its effects may be summarised as follows:

1. It aids the formation of graphite. In this respect, it is approximately one-third as effective as silicon.

2. It has a grain-refining effect. More uniform properties are therefore obtained in castings of varying cross-section since nickel prevents the formation of a coarse grain in thick sections, whilst preventing the formation of a hard white iron in the thinner sections.

3. It lowers the critical cooling rate and the critical range. Progressive increases in nickel content change the microstructure of the matrix from pearlite to martensite to austenite.

	Cast	*Annealed*
Microstructure	PEARLITE + GRAPHITE Above 50 mm thick a pearlite-ferrite matrix is possible	FERRITE + GRAPHITE Annealed at 900°C. for a few hours followed by controlled cooling
Mechanical properties		
TS N/mm²	540–700	420–540*
% Elongation	1–5	10–25
Compressive strength	65–80	45–58
HB	230–280	140–180
Izod Value	4	12

* Annealed S.G. cast iron shows a well-defined yield point on the stress-strain curve.

Table 9.3. S.G. Cast Iron

Nickel cast irons may therefore be classified according to the structure of the matrix, as follows:

1. Pearlitic cast irons, containing up to 2% nickel. These are the most widely used type of nickel cast iron and are used for high-duty castings, particularly those of varying cross-section.

2. Martensitic cast irons. These are hard and wear resisting. The best example is *Ni-Hard* details of which are as follows:

Name	*TC*	*Ni*	*Cr*	*Brinell Hardness**	
				Sand cast	*Chill cast*
Ni-Hard	3·2–3·6	4·5	1·5	550–650	600–725

* Tungsten Carbide Ball

3. Austenitic cast irons. These are all non-magnetic, and possess good heat and corrosion resistance, and high coefficients of expansion and of electrical resistance. Typical examples are:

Trade name	*TC*	*Ni*	*Cr*	*Si*	*Other elements*
'Ni-Resist'	2·8	14	2	1·6	6% Cu
'No-Mag'	2·8	11	–	1	6% Mn
'Nicrosilal'	2·0	18	2	5	–

Table 9.4. Compositions of some Typical Austenitic Alloy Cast Irons

GROWTH OF CAST IRON

When cast iron is heated and cooled through the range 700°–800°C. the cementite in the pearlite breaks down to form ferrite and graphite, thus producing an increase in volume. Subsequently hot gases penetrate into the minute cavities formed and oxidise the ferrite, giving a further increase in volume. This gives rise to stresses which may cause warping and characteristic crazy-cracking at the surface. Irons such as *Nicrosilal* (Table 9.4) have been developed to resist growth.

10. Copper and its Alloys

Copper is widely used industrially due mainly to its high electrical and thermal conductivity, good corrosion resistance and workability. It is employed extensively in the electrical industry for conductors of all kinds. It is also used in chemical plant construction and in the building industry for such applications as domestic water pipes and roofing.

There are many grades of commercial copper available, e.g.

1. Tough-pitch Copper (containing 0·05% oxygen)

When tough-pitch copper is heated above 400°C. in reducing atmospheres it is liable to a defect known as 'gassing'. The hydrogen or carbon monoxide penetrates into the metal and reacts with the cuprous oxide present to form steam. The steam produced is unable to escape and builds up a pressure which forces individual grains apart, forming intercrystalline fissures. The ductility of the metal is considerably reduced. Such conditions may be realised during gas welding, and deoxidised copper is usually specified for welding purposes.

2. Deoxidised Copper

Copper is usually deoxidised with phosphorus which has a higher affinity for oxygen than has copper. Deoxidised copper, although more suitable for welding, usually contains 0·05% phosphorus as a residual deoxidant which reduces the electrical conductivity.

3. O.F.H.C. Copper

Oxygen-free high conductivity (O.F.H.C.) copper contains neither oxygen nor residual deoxidant and therefore possesses a high electrical conductivity. It is made by melting and casting electrolytically refined copper in a non-oxidising atmosphere.

4. Arsenical Copper

0·3–0·5% arsenic may be added to deoxidized or tough-pitch copper to improve the resistance to scaling at elevated temperatures. Arsenical copper is used for domestic plumbing, chemical plant, stay-bolts and rivets.

5. Free-machining Copper

Approximately 0·5% selenium or tellurium may be added to copper to improve machinability. These elements do not greatly decrease the electrical conductivity which is usually about 95% that of pure copper.

The mechanical properties of the various grades of copper do not vary very much. Copper is soft and ductile and capable of a considerable degree of cold-work. Typical mechanical properties in the cold-worked and annealed condition are:

0·1% PS *N/mm²*	*TS* *N/mm²*	*El*	*HV*
62	220	55	45

COPPER ALLOYS

The more important copper-base alloys may be classified as follows:

(*a*) Copper-zinc (Brasses)
(*b*) Copper-tin (Tin bronzes)
(*c*) Copper-tin-Phosphorus (Phosphor-bronzes)
(*d*) Copper-aluminium (Aluminium bronzes)
(*e*) Copper-nickel (Cupro-nickels)

The study of these alloys is to some extent simplified by a similarity in the microstructures. The initial solid solution produced by the addition of an alloying element is represented by the symbol α. The α solid solution is always soft and ductile and has the same type of microstructure in all the alloys. When the limit of solid solubility has been exceeded the β constituent appears in the microstructure, and this structure is associated with increased strength at the expense of ductility. Further addition of alloying element results in the formation of hard brittle constituents represented by the symbols γ and δ.

The Brasses

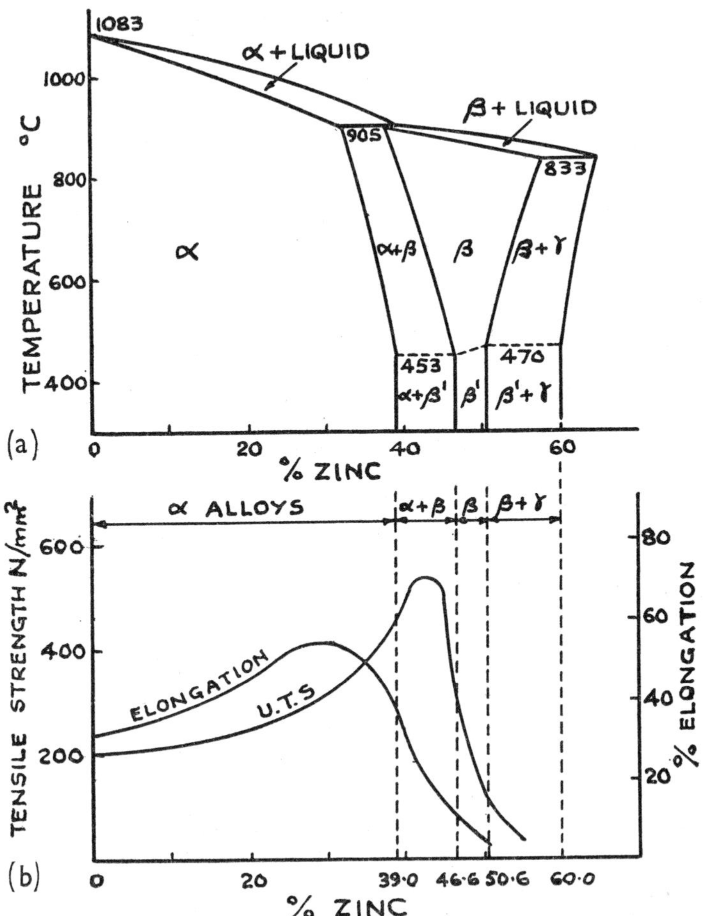

Fig. 10.1. (*a*) The Copper-rich Portion of the Copper-zinc Equilibrium Diagram

(*b*) Graph to show Relationship between Microstructure and Mechanical Properties of the Brasses

Reference to the copper-zinc thermal equilibrium diagram Fig. 10.1(*a*) shows that the soft α solid solution exists in alloys containing 0–39% zinc. Alloys containing 39–46·6% zinc have a microstructure consisting of $\alpha + \beta$, whilst those containing 46·6–50·6% zinc consist entirely of the β constituent. Brasses with more than 50·6% zinc contain the brittle γ constituent and such alloys are not commercially important.

The change from β to β^1 which occurs between 453°–470°C. (Fig. 10.1(*a*)) is not important to the engineer and the symbol β will be used to describe the room temperature β^1 structure.

The relationship between zinc content, microstructure and mechanical properties is shown in Fig. 10.1(*b*). It is apparent that for maximum ductility an alloy containing 30% zinc should be employed. The optimum combination of strength and ductility is obtained in an alloy containing about 40% zinc. These two alloys form the basis of the chief industrial brasses.

1. 70:30 Brass (α brass)

Cold-working brass used for cold-rolled sheets, wire drawing, deep drawing, pressing and in tube manufacture. It is frequently known as *cartridge brass* because of its use in the manufacture of cartridge cases.

In the cast state the alloy consists of a cored structure of the α solid solution. Annealing after cold-work produces a twinned crystal structure (Fig. 10·3(*a*)). This type of structure is characteristic of face-centred cubic alloys in this condition. Typical mechanical properties of 70:30 brass in the cold-worked and annealed condition are as follows:

0·1% PS N/mm²	*TS N/mm²*	*El*	*HV*
77	320	70	65

Improvement in corrosion resistance can be obtained by the addition of tin or aluminium, and alloys such as aluminium brass (76:22:2 Cu:Zn:Al) and Admiralty brass (70:29:1 Cu:Zn:Sn) are used for marine condensers and other heat-exchange equipment.

2. 60:40 Brass ($\alpha\beta$ brass)

Suitable for hot-working by rolling, extrusion and stamping and for the manufacture of castings. This type of brass is frequently known as *Muntz metal* as it was developed by G. F. Muntz.

In the cast state the microstructure is of a Widmanstatten type, Fig. 10.3(*b*), with α at the grain boundaries and along the crystal planes of the β constituent. Hot-working refines the structure giving a more uniform distribution of α in a matrix of β (Fig. 10.3 (*c*)). Typical mechanical properties of 60:40 brass are as follows:

Condition	*0·1% PS N/mm²*	*TS N/mm²*	*El*	*HV*
Cast	90	280–370	25–50	60–70
Hot rolled	90–230	300–460	25–40	75–150

Approximately 2–3½% lead is frequently added to brasses of this type to improve machinability. Lead is insoluble in the brass and appears as small globules in the microstructure which cause the machining chips to break up into small pieces.

3. High-tensile Brasses

These are essentially brasses of the 60:40 type which contain additional elements such as manganese, aluminium, tin, iron and nickel. These 'high-tensile' brasses may be used in the cast or wrought state for applications where strength coupled with a good corrosion resistance is required. Applications include marine propellers, autoclaves, pump rods and shafts, pickling crates, stampings and pressings for automobile fittings and switch-gear. Typical compositions and mechanical properties of high-tensile brasses are shown in the following table:

BS 1400	*Cu min.*	*Mn max.*	*Al max.*	*Fe max.*	*Sn max.*	*Ni Max.*	*Microstructure*	*TS min.*	*El min.*
H.T.B 1	55	3	2·5	0·7-2·0	0·5	1	$\alpha+\beta$*	470	18
H.T.B 3	55	4	3–6	3·25	0·2	1	β	740	11

* Area of α constituent = 15% minimum.

The separate effects of added elements on the properties of brasses are as follows:

Tin improves the corrosion resistance of the brass but should not exceed 1% owing to the formation of a hard brittle constituent. Tin is present in naval brass (61:38:1 Cu:Zn:Sn) for this purpose.

Iron forms a complex bluish iron-rich constituent when present in amounts greater than 0·35%. This constituent provides nuclei for crystallisation and thus has a grain-refining effect in castings.

Manganese is often added up to 2% in the high-tensile brass known as 'manganese bronze'. Its principal effect is that of a deoxidant, producing sounder castings and improving the tensile strength.

Aluminium greatly increases the tensile strength and the corrosion resistance of the brass. It reduces zinc losses during melting by forming a protective film of aluminium oxide on the surface of the molten alloy. This film can, however, provide difficulties in casting and soldering.

Nickel has only a slight effect on mechanical properties and its main function is to increase corrosion and erosion resistance. It is therefore present in manganese-bronze for marine propellers.

The Tin Bronzes

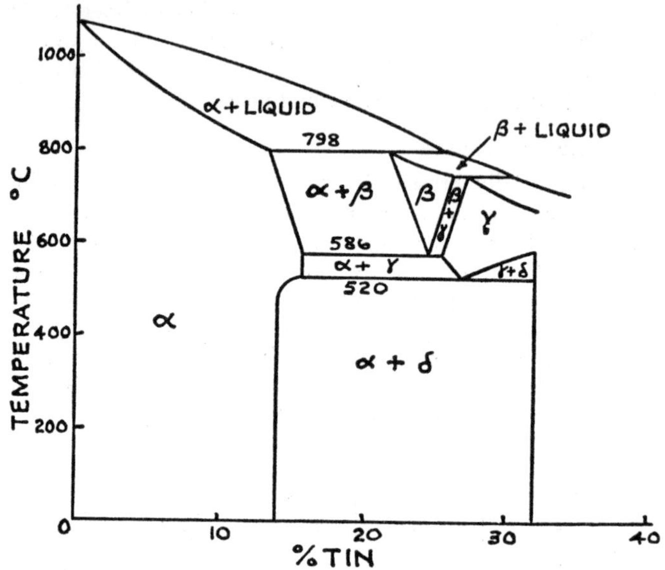

Fig. 10.2. The Copper-rich Portion of the Copper-tin Equilibrium Diagram

Reference to Fig. 10.2 reveals that at room temperatures up to 14% tin can dissolve in copper to form the *a* solid solution. However, in practice the brittle δ constituent appears with 7% tin in sand castings and with 5% in chill castings. The commercial bronzes rarely contain more than 15% tin so that the microstructure of these alloys will consist of either:

(1) Soft *a* constituent alone

or (2) Hard (*a* + δ) eutectoid in a soft *a* matrix. The main types of commercial tin-bronze are as follows:

1. Coinage Bronze (*a* bronze) 95:4:1 Cu:Sn:Zn

The zinc is present mainly as a deoxidiser. The alloy is soft and ductile and in the cold-worked and annealed condition will consist of twinned crystals of the *a* solid solution. The alloy is the standard British 'copper' coinage.

2. Admiralty Gun-metal 88:10:2 Cu:Sn:Zn

Zinc is present as a deoxidiser and also increases the fluidity of the casting. In the cast state the alloy consists of the hard $(\alpha + \delta)$ eutectoid in a cored α matrix (Fig. 10.3(*d*)). Due to the presence of the eutectoid it cannot be cold-worked but may be hot-worked above 590°C. Admiralty gun-metal is used chiefly for castings requiring strength combined with a good corrosion resistance, e.g. high-pressure steam and water fittings. The microstructure of hard particles in a soft matrix gives the alloy good bearing properties. Typical mechanical properties are as follows:

Condition	*0·1% P.S. N/mm²*	*TS N/mm²*	*El*	*HV*
Sand cast	130–160	270–340	12–20	70–100

This is not a good alloy for producing pressure-tight castings but can be improved by the addition of 1% Pb without loss of strength. It is not recommended for use at temperatures above 450°F.

Gun-metals have been developed where lead has replaced the more expensive tin with certain advantages. The 85:5:5:5 Cu:Sn:Zn:Pb alloy (BS 1400 LG2) has moderate strength at room temperatures but possesses excellent pressure tightness. It is superior to other gun-metals for use at temperatures in the region 450°–550°F. The 86:7:5:2: Cu:Sn:Zn:Pb alloy (BS 1400 LG3) has little advantage over the 85:5:5:5 alloy as regards strength and is considerably inferior for pressure tightness.

The addition of nickel to leaded gun-metals has led to the development of the 86:6·5:3:3·5:2 Cu:Sn:Zn:Pb:Ni alloy by International Nickel Ltd. Castings in this leaded gun-metal have mechanical properties equivalent to 88:10:2 combined with the castability and pressure tightness of 85:5:5:5 gun-metal and they are less sensitive to the effect of variations in casting section.

3. 15% Tin-bronze Alloy

The cast 15% tin-bronze is suitable for use as a bearing alloy, since it consists of hard particles of $(\alpha + \delta)$ eutectoid in a soft α matrix. When the alloy is water quenched from about 600°C. the structure consists of $\alpha + \beta$ and in this condition it is used for bells, since the tougher β constituent gives a better tone.

Lead is frequently added to tin-bronzes to improve machinability (0·5–1·0%) and bearing qualities (5–15%). Leaded bronzes for bearings will be discussed in Chapter 12.

(a) 70:30 BRASS, COLD WORKED AND ANNEALED CONDITION. TWINNED CRYSTALS OF THE α SOLID SOLUTION.

(b) 60:40 BRASS, CAST CONDITION. LIGHT FLAKES OF α SOLID SOLUTION IN A MATRIX OF DARK β CONSTITUENTS.

(c) 60:40 BRASS, HOT WORKED. (TRANSVERSE SECTION.) LIGHT ISLANDS OF α SOLID SOLUTION IN A DARK MATRIX OF THE β CONSTITUENT.

(d) ADMIRALTY GUNMETAL 88:10:2. Cu:Sn:Zn. CAST CONDITION. ISLANDS OF $(\alpha+\delta)$ EUTECTOID IN A CORED MATRIX OF THE α SOLID SOLUTION.

Fig. 10.3. Typical Microstructures of Copper-base Alloys

Phosphor-bronze Alloys

The phosphor-bronze alloys are tin-bronzes containing small amounts of phosphorus, consequently they may be studied with reference to the copper-tin equilibrium diagram. Two distinct types of phosphor-bronze may be recognised, viz. (1) *wrought* and (2) *cast phosphor-bronze*, as shown in the table on p. 102.

Aluminium-bronzes

The aluminium-bronze alloys are finding increasing application due to a useful combination of properties. These alloys are capable of heat-treatment in a similar manner to plain carbon steels, and in the wrought heat-treated condition a high tensile strength coupled with good ductility is obtained. Aluminium-bronzes possess good working properties, wear, fatigue and corrosion resistance. Certain difficulties are met, however, in the production of these alloys due to the formation of hard tenacious oxide films on the surface of the molten alloy. Special casting procedures have to be adopted so as to avoid turbulence, and to prevent the oxide entering the casting forming

	Wrought Phosphor-Bronzes	*Cast Phosphor-Bronzes*
Range of composition	3·0–8·5% Sn 0·1–0·3% P	9·0–13·0% Sn 0·3–1·0% P
Microstructure	α In the cold worked and annealed condition twinned crystals are revealed	$\alpha + (\alpha + \delta)$ Plates of the hard constituent Cu_3P (copper phosphide) may be associated with the eutectoid
Typical Mechanical Properties	PS = 130–150 TS = 340–370 % El = 65 HV = 75	PS = 130–150 TS = 220–310 % El = 3–15 HV = 70–110
Uses	Instrument springs Steam turbine blading	Bearings

brittle films. The same difficulties were also obtained in fusion welding, but these have largely been overcome and satisfactory arc welding of aluminium-bronzes is possible.

The portion of the copper-aluminium diagram relevant to the aluminium-bronzes is shown in Fig. 10.4.

Alloys containing up to 9·4% aluminium consist of the soft ductile α solid solution. The γ_2 constituent is hard and brittle and therefore industrial aluminium-bronzes rarely contain more than 10% aluminium.

Two main types of aluminium-bronze are employed:

1. Wrought α Alloy

This contains 5–7% aluminium, and may be readily hot- or cold-worked. In the cold-worked and annealed condition the microstructure will consist of twinned crystals of the α solid solution, similar to Fig. 10.3(a).

The 5% aluminium alloy, due to its colour, is used for imitation jewellery and for decorative purposes, whereas the 7% alloy containing other elements such as nickel, iron and manganese is used for the tubes of marine condensers and other heat exchangers.

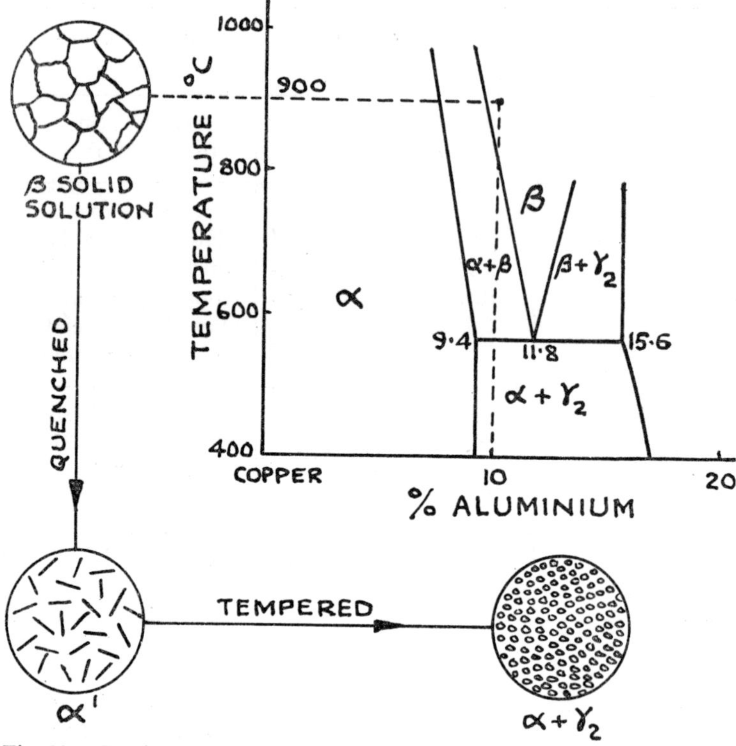

Fig. 10.4. Portion of the Copper-aluminium Diagram illustrating Structural Changes occurring during Heat-treatment of the 10% Aluminium-bronze Alloy

2. 10% Aluminium-bronze Alloy

This is the most important alloy and is used both for castings and in the hot-worked condition. When the alloy is slowly cooled the microstructure consists of dendrites of the α solid solution in a coarse eutectoid matrix ($\alpha + \gamma_2$). The presence of this coarse eutectoid embrittles the alloy and this defect known as 'self-annealing' may be overcome by the addition of 1–3% iron which refines the structure or by heat-treatment. The finer eutectoid obtained in chill cast iron-containing castings resembles the β constituent in other copper-base alloys.

The usual heat-treatment for a 10% aluminium-bronze alloy is as follows:

1. Water quenching from 900°C.
2. Tempering at 550°–650°C.

The effect of heat-treatment on the microstructure of the alloy is shown in Fig. 10.4. The changes taking place are analogous to those occurring in the plain carbon steels. At 900°C. the alloy consists of a uniform solid solution β (analogous to austenite). Water quenching from this temperature results in a hard needle-like constituent referred to as a^1 (analogous to martensite). Tempering at 550°–650°C. results in the precipitation of fine particles to give a tough structure of $a + \gamma_2$ (analogous to sorbite). Typical mechanical properties in the hot-worked and fully heat-treated condition are as follows:

TS *N/mm²*	*El*	*HB*
620–700	30–40	170–180

The 10% aluminium-bronze alloy may also contain additions of iron (up to 5%), nickel (up to 5%) and manganese (up to 2·5%). These complex aluminium-bronzes may be used as castings or in the hot-worked condition. In hot-worked condition a tensile strength of up to 770 N/mm² can be obtained with 15–25% elongation.

Aluminium-bronze alloys are used in the marine and chemical industries for applications where strength coupled with a good corrosion resistance is important, e.g. propellers, propeller shafts, pump castings, pickling crates, chains and hooks. They are being increasingly used as non-sparking tools.

Copper-nickel Alloys

Copper and nickel form a complete series of solid solutions (Fig. 10.5). In the cast state the alloys will exhibit coring which can

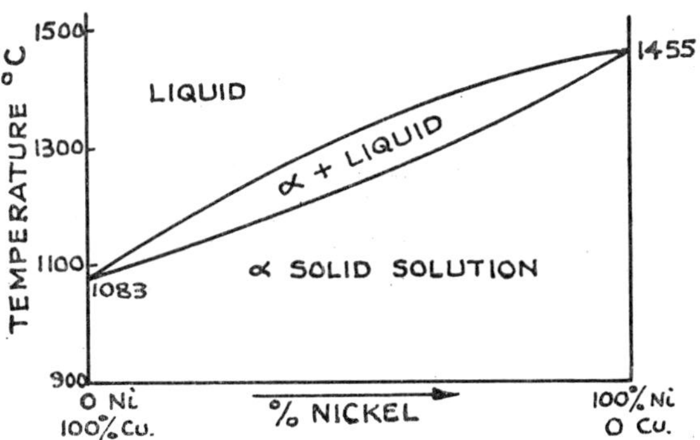

Fig. 10.5. The Copper-nickel Equilibrium Diagram

be removed by annealing. All the alloys may be fabricated by hot- and cold-working. Annealing after cold-working may be carried out between 550°–750°C.

The copper-rich alloys known as the *cupro-nickels* are extremely malleable and are capable of cold-work without intermediate annealing. The 80:20 and 70:30 cupro-nickels are used for condenser tubes, whilst the 75:25 alloy is used for the present-day British 'silver' coinage.

Typical mechanical properties of the cupro-nickel alloys in the annealed condition are as follows:

TS N/mm^2	*El*	*HB*
300–380	40–45	75–80

Copper-beryllium Alloy

The copper-beryllium alloy containing 2% beryllium and up to 0·5% cobalt exhibits the property of precipitation—hardening to a marked degree. This phenomenon will be discussed in more detail in Chapter 11.

Quenching from 800°C. produces a soft supersaturated solid solution. This can be hardened by cold-work, and tempering at 300°–320°C. hardens both the soft and work-hardened alloy.

The combined effect of work hardening and precipitation or temper-hardening gives the following typical mechanical properties:

LP N/mm^2	*TS* N/mm^2	*El*	*HB*
740	1 390	2	365

These alloys are characterised by a high fatigue limit and good elastic properties, and are used for springs, bellows, diaphragms for pressure recording instruments, non-sparking tools, blades for electrical switches, socket connectors and resistance welding electrodes.

11. The Light Alloys

ALUMINIUM

Aluminium is extracted from the mineral bauxite, which consists mainly of aluminium oxide, by an electrolytic process. The process, developed simultaneously by Hall in the U.S.A. and Heroult in France in 1886, involves the electrolysis of a solution of bauxite in fused cryolite. The purity of the aluminium produced is about 99·5%, but this can be raised to about 99·99% by further refining.

One of the most important characteristics of aluminium is its lightness. The relative density is 2 710 compared with 7 800 for mild steel. A high strength:weight ratio is therefore obtained in heat-treated aluminium alloys which makes them suitable for use in the aircraft and automobile industries. The thermal conductivity of aluminium is about five times that of mild steel and this property, together with that of lightness, makes the metal suitable for pistons and connecting rods of internal combustion engines. The electrical conductivity of aluminium is about 60% that of copper, but weight for weight it is better than copper. It is therefore used for cable work, provided a central core of thin steel wire is used as reinforcement. Aluminium has a high affinity for oxygen and readily forms a thin hard self-healing film of aluminium oxide on its surface. This protective surface film, which is only about 1×10^{-5} mm thick, accounts for the good corrosion resistance of aluminium. It is particularly resistant to concentrated nitric acid, but not to alkalies, which dissolve the film. Aluminium and its alloys are therefore used in chemical plant construction, marine superstructures, food containers and wrappers, cooking utensils and aluminium paints.

Commercially pure aluminium is soft and weak. In the annealed state it exhibits a tensile strength of about 80 N/mm^2, elongation of 35% and Brinell hardness of 23. It is therefore unsuitable for most engineering purposes. The mechanical properties can be improved by alloying, the chief alloying elements being copper, silicon, manganese, magnesium, zinc and nickel.

ALUMINIUM ALLOYS

Aluminium alloys are available under a considerable number of trade names which can be very confusing. For general engineering purposes aluminium alloys are covered by British Standards 1470–75 (wrought forms) and 1490 (ingots and castings). Aircraft materials are covered by the 'L' series of British Standards and the D.T.D. Specifications. The latter are issued by the Directorate of Technical Development. Aluminium alloys may be conveniently classified as follows:

1. Wrought Alloys (*a*) not heat-treated
   (*b*) heat-treated
2. Cast Alloys (*a*) not heat-treated
   (*b*) heat-treated

1. Wrought Alloys

(a) Not Heat-treated

The chief alloys in this class are the 1¼% manganese alloy and those containing 2–7% magnesium. They are strengthened only by cold-working and are available in various tempers, e.g. 'soft', 'half hard', 'three-quarter hard', etc. Softening may be carried out by annealing at 350°–400°C. These alloys are characterised by good corrosion resistance. The magnesium-containing alloys are particularly resistant to sea-water corrosion, and are therefore used for many ship-building applications. The 1¼% manganese alloy is used extensively for roofing sheets, wall cladding for various buildings and food containers. Typical mechanical properties for a 5% magnesium alloy are as follows:

Composition	*Form*	*Condition*	*0·1% PS* N/mm^2	*TS* N/mm^2	*El*
5% Mg	Sheet	Soft	110	260	20
		Half hard	280	320	6

(b) Heat-treated

The alloys in this group are capable of being hardened and strengthened by a process known as 'age-hardening'. The usual alloying elements are copper, magnesium, manganese, silicon and zinc. The phenomenon of age-hardening will be discussed with reference to alloys containing 4% copper, which is the basic composition of alloys of the Duralumin type. The aluminium-rich portion of the aluminium-copper diagram is shown in Fig. 11.1.

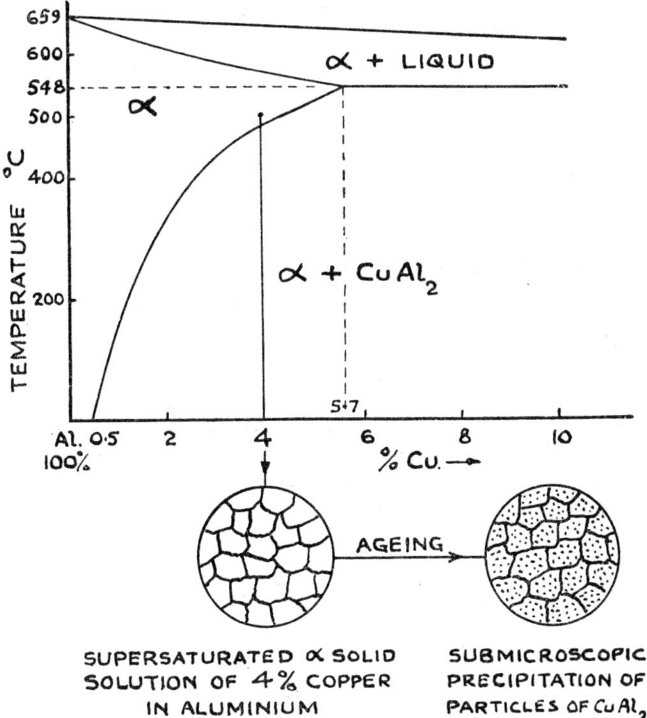

Fig. 11.1. Aluminium-rich Portion of Aluminium-copper Equilibrium Diagram illustrating the Age-hardening Treatment of an Alloy containing 4% Copper

At room temperatures the solubility of copper in aluminium to form the *a* solid solution is only 0·5%. The solubility increases with temperature reaching a maximum of 5·7% at 548°C. Age-hardening is a characteristic of alloys with a falling solid solubility curve of this type.

If an alloy containing 4% copper is cooled slowly from the *a* region, coarse particles of $CuAl_2$ will be precipitated, due to the decrease in solubility of copper in aluminium with fall in temperature. In this condition the alloy is relatively weak and brittle.

Improved mechanical properties are obtained by the full age-hardening treatment which involves three distinct stages:

1. Heating the alloy to a prescribed temperature (e.g. 500°C.) to dissolve the copper and any other alloying elements.
2. Quenching from this temperature to preserve the α solid solution at room temperatures. This rapid cooling prevents the precipitation of the copper as $CuAl_2$, and a supersaturated solid solution of 4% copper in aluminium is obtained. These two treatments are referred to as *Solution Treatment.*
3. If the alloy is kept at room temperature hardening now occurs spontaneously, when it is referred to as age-hardening or natural-ageing. This is due to the sub-microscopic precipitation of particles of $CuAl_2$ and other compounds such as Mg_2Si. These particles act as obstacles to the passage of dislocations thereby increasing the hardness. With natural-ageing the maximum hardness is reached after about four days.

Natural-ageing may be delayed by storing the work in a refrigerator at $-6°$ to $-10°C$. This is important where riveting and pressing operations are employed and delays are likely to occur.

Hardening can be accelerated by reheating the quenched alloy to approximately 150–170°C. for a few hours. It is then referred to as artificial-ageing or precipitation-hardening. The time and temperature of ageing differ with each alloy and must be closely controlled to give optimum results (Fig. 11.2).

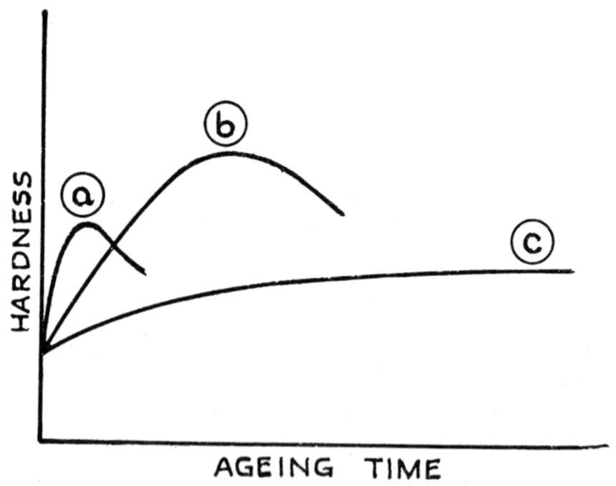

Fig. 11.2. The Effect of Ageing Temperature and Time on the Hardness of a Duralumin-type Alloy

(*a*) Temperature too high
(*b*) Optimum temperature
(*c*) Room temperature

In artificial ageing the hardness reaches a maximum more quickly and then falls off. The fall in hardness is due to the coalescence of the precipitate into fewer and larger particles, thus providing less obstacles to the passage of dislocations. The maximum hardness is reached before the precipitate becomes visible under the microscope.

The original Duralumin alloy has an approximate composition of 4% copper with 0·5% each of magnesium, manganese, silicon and iron. The last two elements are strictly impurities in the alloy. The heat-treatment of this alloy involves solution treatment at 490°–500°C. followed by natural ageing for four days. In the fully aged condition typical mechanical properties are as follows:

0·1% PS N/mm²	*TS N/mm²*	*El*	*HB*
230	390	15	110

The alloy is used in the form of bars, tubes, sheet, forgings, rivets for general purposes and stressed parts in aircraft.

The addition of controlled amounts of silicon up to a maximum of 0·9% produces a higher strength Duralumin-type alloy (H.15) used extensively for highly stressed parts in aircraft. Typical mechanical properties are as follows:

0·1% PS N/mm²	*TS N/mm²*	*El*	*HB*
400	460	8	135

These properties are obtained by solution treatment at 510°C. followed by precipitation hardening at 170°C. for about ten hours.

The highest strength alloys are those of the Al-Zn-Mg-Cu type, e.g. D.T.D. 683. A typical composition is approximately 6% Zn, 2·5% Mg and 1·5% Cu. After precipitation-hardening such an alloy would possess a tensile strength of 540–590 N/mm².

The Al-Mg-Si alloys which contain approximately 1% Mg and 1% Si, but no copper, have a high corrosion resistance but are not as strong as the copper-containing Duralumin alloys.

The wrought heat-treated alloys are frequently protected from corrosion by a layer of pure metal (*Alclad* or *Aldural*). This pure metal cladding is applied by a hot-rolling process (page 146). Clad alloys therefore combine high strength with good corrosion resistance.

2. Cast Alloys

(a) Not Heat-treated

The most important casting alloys are those of the aluminium-silicon type, containing 10–13% silicon. These alloys are characterised by low specific gravity, low shrinkage, high fluidity and pressure tightness. They are suitable for sand castings, and gravity and pressure die castings. The aluminium-rich portion of the aluminium-silicon diagram is shown in Fig. 11.3.

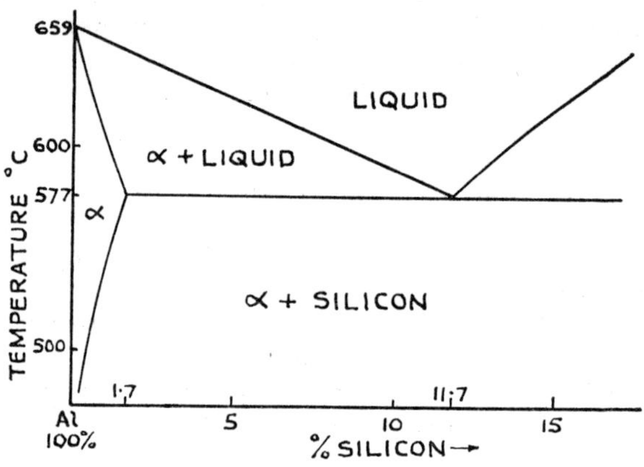

Fig. 11.3. The Aluminium-rich Portion of the Aluminium-silicon Equilibrium Diagram

It will be observed that these alloys are of approximate eutectic composition. The normal mechanical properties of these alloys are greatly improved by a process known as *modification*. This involves the addition of approximately 0·05% sodium to the molten alloy. A considerable refinement of the eutectic structure results with consequent improvement in mechanical properties. Modification can increase the tensile strength from 120 to 200 N/mm² and the percentage elongation from 5 to 15%. Typical mechanical properties are as follows:

BS 1490	*Si%*	*Condition*	*0·1% PS N/mm²*	*TS N/mm²*	*El*	*HB*
LM 6	10–13	Sand Cast	60	180	5–10	50–55
		Chill cast	70	190–230	7–15	55–60

Alloys of this type are known by various trade names, e.g. *Alpax* and *Silumen*.

These alloys possess good corrosion resistance and are suitable for marine castings, automobile fittings, water-cooled manifolds and jackets, thin-section and intricate castings such as motor housings, meter cases and switch boxes, and castings for the dye and chemical industries.

Casting alloys of the Al-Mg-Mn type containing 3–6% Mg and 0·3–0·7% Mn are noted for their high corrosion resistance particularly in marine atmospheres. Such alloys are suitable for sand and gravity die castings. Pressure die castings are possible only in relatively simple shapes.

(b) Heat-treated

The heat-treatment of cast alloys is similar to that of the wrought alloys although the times required are generally longer.

The earliest alloy of this type was Y alloy developed during the 1914–18 war as an aero-engine piston alloy. Y alloy contains approximately 4% Cu, 2% Ni and 1·5% Mg. The full heat-treatment involves solution treatment at 500°–520°C for six hours followed by natural ageing for five days or precipitation-hardening for two hours at 100°C. Typical mechanical properties in the chill-cast and fully heat-treated condition are as follows:

0·1% PS N/mm^2	*TS* N/mm^2	*El*	*HB*
220–250	280–310	1–3	100–130

Y alloy can also be used in the form of forgings, but is chiefly used in the cast form. It retains its strength at elevated temperatures and is used for diesel and high-duty petrol engine pistons and cylinder heads.

The alloy *Lo-Ex* corresponding to BS 1490 LM13 is noted for its low thermal expansion and is used for pistons for all types of diesel and petrol engines, and other engine parts operating at elevated temperatures. This alloy is of the 11–13% silicon type and contains approximately 0·9% Cu, 1·2% Mg and 2·5% Ni.

The alloy RR50 is used for sand and gravity die castings for aircraft and road vehicle engines. This alloy can be precipitation-hardened at 150°–175°C for 8–24 hours without preliminary solution treatment. A typical composition for RR50 together with mechanical properties is as follows:

Cu	*Ni*	*Mg*	*Fe*	*Ti*	*Si*	*0·1% PS* N/mm^2	*TS* N/mm^2	*El*
1·5	1·2	0.15	1·0	0·15	2·5	120	200	3

MAGNESIUM

The development and application of magnesium alloys have increased considerably over the past twenty years. Pure magnesium has poor mechanical properties and is not suitable for engineering applications. This can be overcome by alloying and subsequent working or heat-treatment. The characteristic properties of magnesium and its alloys are as follows:

1. Lightness

The relative density of magnesium is 1 740 compared with 2 710 for aluminium and 7 800 for steel. Magnesium alloys suitably heat-treated possess a high strength-to-weight ratio.

2. High Affinity for Oxygen and Nitrogen in the Molten Condition

Special fluxes are necessary to avoid the formation of oxide and nitride inclusions in the metal.

3. Good Machinability

Magnesium alloys can be machined faster than other metals, but certain elementary precautions are necessary to eliminate fire risk.

4. Workability

Magnesium has a hexagonal close-packed lattice structure. Metals possessing this structure are not readily cold-worked. Magnesium alloys, particularly those containing zirconium, are readily hot-worked in the range 300°–500°C.

5. Corrosion Resistance

The resistance to atmospheric corrosion may be considered slightly better than that of mild steel, but protection is usually given by paint on a chromated surface (page 147). A marked improvement in corrosion resistance is obtained by reducing the iron and nickel contents. The 1½% manganese alloy is noted for its good corrosion resistance.

Classification of Magnesium Alloys

Magnesium alloys may be conveniently classified as follows:

(1) Magnesium-manganese alloys
(2) Magnesium-aluminium-zinc alloys
(3) Zirconium-containing alloys.

Magnesium alloys are covered by BS 2970 and BS 3370-74. In this discussion the alloys will be given the appropriate *Elektron* alloy specification.

1. Magnesium-manganese Alloys

E.g. A.M. 503 containing 1·5% Mn. These alloys are used in the wrought condition and possess good corrosion-resistance and weldability. Typical mechanical properties in the form of sheet are as follows:

0·1% PS *N/mm²*	*TS* *N/mm²*	*El*
80–160	200–280	5–14

2. Magnesium-aluminium-zinc Alloys

These are used in both the cast and wrought forms. Typical examples are as follows:

Elektron Alloy	*Composition*		*Condition*	*0·1% PS* *N/mm²*	*TS* *N/mm²*	*El*
	Al	*Zn*				
AZ 91	9·5	0·5	Cast + heat-treated	100–130	200–260	1–4
AZM	6·0	1·0	Extruded bars section 75 mm	170–220	260–340	10–18

3. Zirconium-containing Alloys

The commercial production of zirconium-containing alloys started in 1946. The addition of 0·6–0·7% zirconium gives rise to pronounced grain refinement and improved values of the 0·1% proof stress. The hot-workability of these alloys is particularly good. Zirconium-containing alloys can be used in both the cast and the wrought conditions. Typical examples are as follows:

Elektron Alloy	*Composition*		*Condition*	*0·1% PS N/mm²*	*TS N/mm²*	*El*
	Zn	*Zr*				
Z5Z	4·5	0·7	Cast + heat-treated	130–160	230–280	5–12
ZW3	3·0	0·7	Press forgings	200–230	290–340	8–14

The addition of thorium and rare-earth elements such as cerium to zirconium-containing alloys greatly improves the creep resistance at elevated temperatures. Examples are ZT1 (3% Th, 2·2% Zn, 0·7% Zr) and ZRE1 (2·7% rare-earths, 2·2% Zn, 0·6% Zr). ZRE1 is creep resistant up to 250°C and possesses excellent casting properties. ZT1 is creep resistant up to 350°C. More recently silver has been added to such alloys. MSR–A (1·7% rare earths, 2·5% Ag, 0·6% Zr) has the highest proof strength of any cast magnesium alloy (150–210 N/mm²). The tensile properties at elevated temperatures as well as the creep properties are also good and the alloy is not liable to stress corrosion.

Magnesium-lithium alloys which contain 14% lithium and 1% aluminium have been introduced recently for use in missiles and other aerospace vehicles. Such alloys are about 22% lighter than pure magnesium.

Applications of Magnesium Alloys

Magnesium castings are used in the aircraft industry for landing wheels, undercarriage legs, gas-turbine engine air intakes and engine support plates and frames. In the automobile industry crankcases and clutch and gear housings may be made from magnesium alloys. The Mg-Al-Zn alloys are used extensively for textile machinery, portable tools, vacuum cleaners, printing machinery and camera bodies. Magnesium is used for the manufacture of anodes for the cathodic protection of steel.

Magnesium alloys in the form of sheet, pressings and forgings are used for air frames and welded petrol and oil tanks and for parts of the fuselage and wings of many aircraft. Magnesium-thorium alloys are widely used in the U.S.A. in rockets, guided missiles, probes and satellites.

About 30–40% of the total magnesium production is used for alloying with aluminium to give the corrosion-resistant aluminium-magnesium alloys.

12. Miscellaneous Non-ferrous Metals and Alloys

NICKEL AND ITS ALLOYS

Nickel

Commercially pure nickel contains approximately 99·5% nickel together with small amounts of Cu, Fe, Mn, Si, C and S. In general with more than 0·005% sulphur a brittle grain-boundary film of nickel sulphide is formed. This can be overcome by the presence of up to 0·2% magnesium to ensure the formation of magnesium sulphide instead of nickel sulphide.

Nickel possesses a good combination of strength and corrosion resistance. It has a particularly good resistance to corrosion by caustic alkalis, ammonia salt solutions and organic acids. It is strongly magnetic.

Typical mechanical properties of commercial nickel in the form of sheet and strip are as follows:

Condition	*0·2% PS* N/mm^2	*TS* N/mm^2	*El*	*HV*
Cold-rolled (hard)	530–590	620–680	8–12	180–210
Cold-rolled + annealed	80–150	340–560	50–35	90–120

Nickel is used as anodes for nickel-plating, for chemical plant construction, and in the manufacture of food-handling equipment, due to its non-toxic properties.

Nickel Alloys

A considerable number of nickel alloys are available for a wide range of applications. Typical well-known alloys manufactured by Henry Wiggin Co. Ltd. are *Monel, Inconel,* the *Nimonic Series,* the *Bright Ray Series* and *Corronel.*

Monel (Monel alloy 500)

Monel contains approximately two-thirds of nickel and one-third of copper with smaller amounts of manganese, iron, silicon and carbon. It has a good resistance to corrosion to fresh and salt water, alkalis, reducing acids, alkaline solutions and super-heated steam. Its good mechanical properties are well maintained at elevated temperatures. Typical mechanical properties in the form of sheet or strip are as follows:

Condition	*0·2% PS*	*TS*	*El*	*HV*
Hard-rolled	590–740	700–830	15–2	200 min.
Cold-rolled + annealed	170–310	480–590	50–30	100–140

Monel is used in both the cast and wrought conditions. It is used in power plant for such applications as valve seats and spindles, pump rods, pump spindles and impellers, corrugated joint rings, needles and nozzles in Pelton wheel installations and turbine blading. It is also used for pickling crates and for chemical and food-processing equipment.

K-Monel (Monel alloy K-500)

The addition of 2–4% aluminium to Monel enables the alloy to be heat-treated so as to obtain improved mechanical properties, whilst retaining the corrosion resistance of Monel. This alloy known as K-Monel also possesses good elastic properties at low and high temperatures.

Solution treatment is carried out at 950°–1 000°C. The alloy is then precipitation-hardened at 590°C. for up to sixteen hours followed by controlled cooling. Precipitation-hardening may be carried out on the soft or cold-worked alloy. In the latter case the effect of work-hardening and temper-hardening may be combined. Typical mechanical properties of K-Monel in the cold-worked and thermally hardened condition are as follows:

0·2% PS *N/mm²*	*TS* *N/mm²*	*El*	*HV*	*Izod* *Joules*
680–900	960–1 160	30–15	280–340	35

K-Monel is used for propeller shafts, highly stressed nuts and bolts, pressure-sensitive instruments, e.g. pressure gauges, bellows and diaphragms.

Inconel (Inconel alloy 600)

Inconel contains approximately 15% chromium, 8% iron, balance nickel. It combines good corrosion resistance with good mechanical properties and resistance to oxidation at elevated temperatures. At elevated temperatures a closely adherent oxide film is formed which retards further oxidation. Typical mechanical properties in the form of sheet and strip are as follows:

Condition	*0·2% PS N/mm²*	*TS N/mm²*	*El*	*HV*
Hard-rolled	540–770	700–930	12–2	230–270
Cold-rolled + annealed	310–390	560–710	30–20	160–200

Inconel is used extensively for food-processing equipment and chemical plant. Due to its heat resistance it is used for aero-engine exhaust manifolds, thermocouple sheaths, protective sheathing for electric-heating elements, furnace components, enamelling racks and garter springs on steam turbines.

The Nimonic Series of Alloys

These alloys are basically nickel-chromium alloys which possess good creep, fatigue and oxidation resistance at elevated temperatures. The high creep strength of such alloys is obtained by the addition of such elements as titanium and aluminium which give rise to precipitation-hardening systems. The main use is in jet engines for flame tubes, rotor blades and discs.

Nimonic 75 is of the 80:20 Ni:Cr type, containing 0·2–0·6% Ti and up to 0·15% C. It is the standard material for gas-turbine flame tubes. Nimonic 80 and 80A are similar in basic composition but contain 0·5–1·8% aluminium in addition to 1·8–2·7% titanium and up to 0·1% carbon. Nimonic 80, introduced in 1941, was the first of these Nimonic alloys developed for gas-turbine blades and discs. Nimonic 80 and 80A are solution treated at 1 080°C. for eight hours, followed by air cooling and aged at 700°C. for sixteen hours, followed by air cooling. These alloys are suitable for use up to 750°C. Nimonic 90 and 95 contain 15–21% cobalt replacing nickel and may be used up to 870°C. and 940°C. respectively. Nimonic 100 is somewhat similar to Nimonic 90 in composition but contains 5% molybdenum, and is used for service temperatures up to 950°C. Nimonic 105 and the latest alloy Nimonic 115 show a further improvement in properties and operating temperature.

It is obvious that alloys which possess strength at high temperatures will present difficulties in hot-working. Many of these difficulties have been overcome by the development of glass lubrication for extrusion.

The Electrical-resistance Alloys

These are based upon the 80:20 Ni:Cr composition and combine high electrical resistance and resistance to oxidation at elevated temperatures up to 1 150°–1 250°C. The Ni:Cr:Fe alloys containing approximately 20% iron are suitable for use up to 950°C. These alloys are known by various trade names such as the 'Bright Ray Series', 'Nichrome' and 'Pyromic'.

Corronel

Corronel 210 contains approximately 66% nickel, 28% molybdenum and 6% iron. The alloy is noted particularly for its resistance to hydrochloric, sulphuric and phosphoric acids.

BEARING METALS

Requirements of a Bearing Alloy

Before an alloy can be used as a bearing metal it must satisfy a number of requirements. It should be sufficiently hard to resist wear and abrasion, tough to withstand shock loads, and strong enough to support the dead weight of the shaft. At the same time it should possess sufficient plasticity to allow for self-alignment. This combination of properties is, with certain exceptions, best achieved by using an alloy with a microstructure consisting of hard particles in a soft matrix. The hard particles provide the necessary wear resistance whilst the soft matrix allows local yielding. In service the soft matrix wears away below the level of the hard constituent. This not only reduces friction but also provides a series of minute oil grooves which aid lubrication.

In addition to the above properties a bearing alloy should possess good melting and casting properties and should bond readily to the backing or support. It is important that the alloy should have a low coefficient of friction to keep the bearing cool. The mechanical properties of the bearing should be maintained at slightly elevated temperatures. A resistance to corrosion by lubricants is also necessary.

A number of alloys are suitable for use as bearings and the choice will depend upon the load upon the bearing and the speed of rotation of the shaft. Bearing alloys may be classified according to composition as follows:

(1) Copper-base bearing alloys
(2) White-metal bearing alloys
(3) Cadmium-base bearing alloys

1. Copper-base Bearing Alloys

The tin bronzes containing 10–15% tin and the cast phosphor-bronzes have already been discussed (Chapter 10). The hard constituent is provided by the ($\alpha + \delta$) eutectoid in a soft matrix of the α solid solution. Other useful copper-base bearing alloys are the leaded bronzes containing 10–15% Pb or the straight copper-lead alloys containing 25–30% Pb. Lead is insoluble in copper and appears as globules in the microstructure. Lead improves plasticity and so allows for any lack of fit or alignment of bearings. The addition of 1–3% tin to the copper-lead bearing alloy assists casting by reducing the tendency to segregation of the lead globules. In general copper-base alloys are capable of carrying higher loads than the white metal bearings.

Porous-bronze bearings can be manufactured by powder metallurgy methods. Such bearings are manufactured by heating a compressed mixture of powdered copper, tin and graphite at about 700°–800°C. The porous-bronze bearing produced behaves like a solid sponge and soaks up considerable quantities of oil. During service this oil is fed to the bearing surface, being filtered as it passes through the metal. The presence of graphite also aids lubrication.

2. White-metal Bearing Alloys

The term ‘white metal’ covers both tin-base and lead-base bearing alloys.

Tin-base Bearing Alloys

The chief elements added to tin to provide the hard constituents are antimony and copper. These Sn-Sb-Cu alloys are frequently referred to as ‘Babbitt metal’ although the original ‘Babbitt metal’ was one specific alloy containing 89% tin, 7·4% antimony and 3·6% copper.

The tin-rich portion of the tin-antimony diagram is shown in Fig. 12.1.

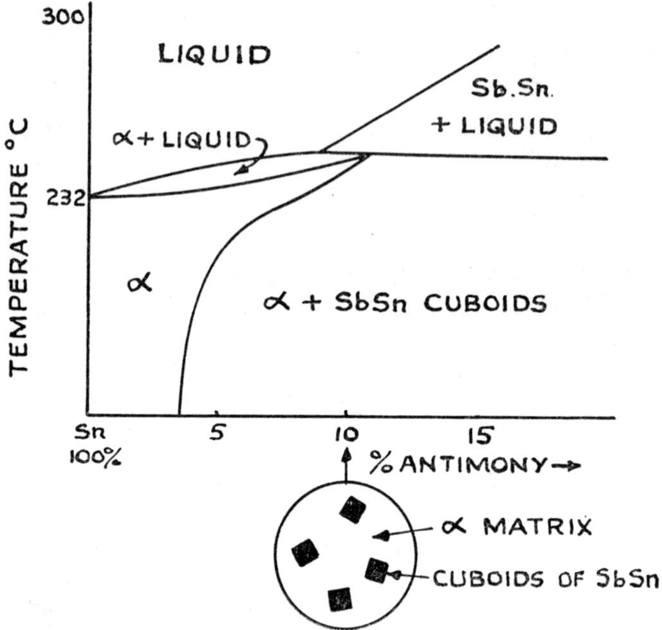

Fig. 12.1. The Tin-rich Portion of the Sn-Sb Equilibrium Diagram illustrating Microstructure of a 10% Sb Alloy

An alloy containing 10% antimony will consist of relatively hard cuboids of the compound SbSn in a soft matrix of the a solid solution. During the freezing of the alloy these cuboids, being lighter than the liquid, tend to float to the surface. The addition of copper gives rise to another hard constituent Cu_6Sn_5 which exists in the form of needle-shaped crystals. These crystals have a higher freezing point than SbSn and, during freezing, they form a network which prevents the SbSn cuboids floating to the surface. The

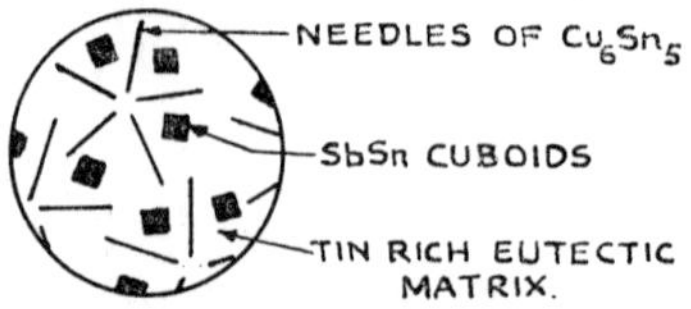

Fig. 12.2. Microstructure of Tin-base Bearing Alloy of Composition 86:10·5:3·5 Sn:Sb:Cu

microstructure of a typical tin-base bearing alloy is shown in Fig. 12.2.

For big-end bearings an alloy containing 93% tin, 3·5% antimony and 3·5% copper is used. This has a Brinell hardness of about 25. The alloy containing 86% tin, 10·5% antimony and 3·5% copper is suitable for main bearings and high-duty bearings for general purposes and has a Brinell Hardness of about 33.

Tin-base bearing alloys are superior in most respects to lead-base bearings, but the latter are cheaper.

Lead-base Bearing Alloys

Lead-base bearing alloys are used for less severe service conditions. Antimony and tin are the chief hardening elements, although alkaline earth metals such as calcium and barium may also be used.

Lead and antimony form a eutectic containing 13% antimony which melts at 248°C. The microstructures of lead-base bearings will vary with composition. Magnolia metal (80Pb 15Sb 5Sn) consists of cuboids of practically pure antimony in a lead-rich eutectic matrix.

3. Cadmium-base Bearing Alloys

Cadmium-base bearing alloys have superior mechanical properties to tin-base alloys, but are inferior to the latter in respect of corrosion resistance, plasticity and anti-frictional properties. Nickel, copper and silver may be present in amounts up to 2%. Hard micro-constituents such as $NiCd_7$ and $CuCd_3$ may be formed, but silver usually goes into solid solution.

ZINC AND ITS ALLOYS

Zinc is mainly used as a protective coating for steel and may be applied by either galvanising, sherardising or spraying (Chapter 14).

Zinc-base alloys are being increasingly used in the form of gravity and pressure die castings. These zinc-base die castings are usually known under the trade name of 'Mazak' alloys and are covered by BS 1004. Typical compositions and properties are as follows:

Alloy	*Nominal composition*			*Original mechanical properties*		
	Al	*Cu*	*Mg*	*TS*	*El*	*HB*
Mazak 3	4·1	—	0·04	280	15·2	83
Mazak 5	4·1	1·0	0·04	330	9·2	92

These alloys undergo a very slight shrinkage on ageing at room temperatures. This shrinkage can be brought almost to completion by a stabilising treatment at 100°C. for six hours, without any appreciable effect on the mechanical properties. Mazak 3 is used where the highest dimensional stability is required and where castings are likely to be subject to heat in service. This alloy is preferred for general-purpose die casting. Mazak 5 is used where maximum castability in production is desired or where a harder and stronger alloy is required.

Very small amounts of tin, cadmium and lead are detrimental since they cause intercrystalline corrosion, causing an increase in dimensions and in some cases excessive swelling and cracking. This has now been overcome by the use of high purity 99·99% zinc.

LEAD AND ITS ALLOYS

Lead was one of the earliest metals known to man. It is mentioned in the Old Testament where it was used for ornamental objects and structural purposes. Lead mines were worked in Britain as early as the first century A.D. It was in general use for water pipes in Roman times.

Lead is characterised by high resistance to corrosion, low melting point (327°C.), high specific gravity (11·3), and good malleability. It is the softest of metals in common use and can be extruded as pipe or rolled into thin sheets. Lead is used for electric cable sheathing, water pipes, roofing sheets, and in the chemical industry for the storage and transport of corrosive liquids.

Lead-base alloys hardened by antimony and tin are used for bearing metals (page 123) whilst lead-tin alloys are used for soft soldering (page 126). Lead-antimony alloys containing 7–12% antimony are used for storage-battery grids.

Type Metals

The ternary alloys of lead, tin and antimony are used for the founding of type for printing purposes. Lead-base alloys used for slug-casting machines such as the Linotype and Intertype machines contain 10–13% antimony and 2–4% tin. Other examples are Monotype alloy (7–10% Sn 14–19% Sb) and Stereotype alloy (3–10% Sn 14–17% Sb). These alloys are characterised by low melting point, high fluidity, good wear resistance and absence of solidification shrinkage.

Fusible Alloys

Fusible alloys are low-melting-point alloys containing bismuth, lead and tin as the chief ingredients. Other metals such as cadmium, antimony, and mercury may also be present. These alloys are usually of eutectic or near-eutectic composition and are used for metal patterns, die mounting, fillers for tube bending, constant-temperature baths for the heat treatment of steel, safety plugs in boilers, and fire sprinklers. The mercury-containing alloys are used in dental work. Some typical fusible alloys are given in the table below:

Name	*Composition per cent.*					*Melting range °C.*
	Bi	*Pb*	*Sn*	*Cd*	*Sb*	
'Cerrobend'	50	26·7	13·3	10	—	70–73
'Cerromatrix'	48	28·5	14·5	—	9	103–227
Wood's Metal	50	25·0	12·5	12·5	—	70–72
Rose's Alloy	50	28·0	22·0	—	—	96–110

13. The Joining of Metals and Alloys

The chief methods used for the joining of metals may be classified as follows:

1. Soft soldering
2. Brazing
3. Welding

SOFT SOLDERING

In soft soldering, as in brazing, a thin film of molten alloy is introduced between the parts to be joined at a temperature below the melting point of those parts. The soft solder employed must fulfil the following requirements:

(*a*) Its melting point must be lower than that of the metals to be joined, but higher than the expected service temperature.

(*b*) It must 'wet' and flow freely over the surfaces of the metal to be joined.

(*c*) It should solidify as a sound film of metal and adhere firmly to the work.

(*d*) It must have adequate mechanical strength. For most soft solders the strength of the joint is about 50–60 N/mm^2.

The soft solders commercially employed are either tin-lead alloys or 'antimonial' tin-lead alloys containing up to 3% antimony. These alloys are covered by BS 219.1959. Antimonial solders are slightly cheaper since antimony replaces expensive tin. They are suitable for most purposes but should not be used for soldering galvanised iron components since they form weak alloys with zinc.

The two main soft soldering alloys are tinman's solder (62% Sn 38% Pb) and plumber's solder (67% Pb 33% Sn) (Fig. 13.1). Tinman's solder is of eutectic composition and is therefore the lowest-melting-point alloy in the series. It is used for electrical, radio and instrument assemblies and machine soldering of can-end seams. Plumber's solder freezes over a range of temperature and thus has a pasty stage which enables the joint to be 'wiped'. Plumber's solder is used in the wiping of lead cable and pipe joints.

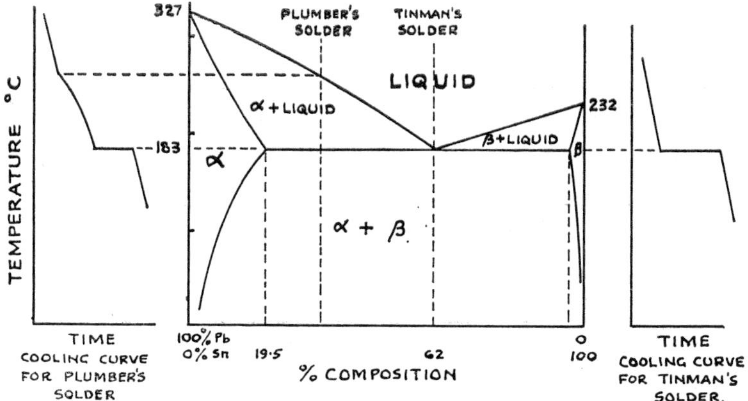

Fig. 13.1. Lead-tin Thermal Equilibrium Diagram showing Cooling Curves for Tinman's and Plumber's Solder

A flux is essential in soft soldering to remove oxide films and expose clean surfaces over which the molten solder can flow. The flux also prevents oxidation of the solder. Typical fluxes are zinc chloride ('killed spirits') and ammonium chloride ('sal ammoniac') or mixtures of both. These fluxes are corrosive and should be removed by washing after soldering. Rosin-cored solder wire (BS 441.1954) is widely used in this country, particularly for electrical purposes. These contain wood or gum rosin with or without an 'activator' which acts as a non-corrosive flux.

BRAZING

From the metallurgical point of view brazing is similar to soft soldering except that a higher-melting-point alloy is used for making the joint. Brazing alloys are stronger and are capable of withstanding higher service temperatures.

Brazing alloys may be classified into eight main types as specified in BS 1845.1966. Three groups will be discussed.

(1) Silver brazing alloys or 'silver solders'
(2) Copper-phosphorus brazing alloys
(3) Brazing brasses.

1. Silver Brazing Alloys

Brazing using these alloys is usually referred to as silver soldering. Silver solders are essentially Ag-Cu-Zn alloys with or without cadmium, which have a lower melting point than the usual brazing brasses. The composition and approximate freezing ranges of these alloys are shown in Table 13.1.

BS 1845 Type	*Ag*	*Cu*	*Zn*	*Cd*	*Approx. Freezing Range °C.*
AG 1	49–51	14–16	15–17	18–20	620–640
AG 4	60–62	27·5–29·5	9–11	—	690–735
AG 5	42–44	36–38	18·5–20·5	—	700–775

Table 13.1. Compositions of Typical Silver Solders

The first of these alloys (AG 1) is known commercially as *Easy-Flo.* It has a small freezing range and its composition approaches that of the Ag-Cu-Zn-Cd eutectic. Silver solders are very free-flowing and strong joints (340–450 N/mm^2) can be made with very little heat-effect on the parent metal.

The choice of flux in silver soldering and brazing generally depends upon the temperature. In general for brazing alloys melting above 760°C. a borax-type flux is used. Below this temperature a fluoride-type flux is employed since those of the borax type are too viscous at these temperatures.

2. Copper-phosphorus Brazing Alloys

The phosphorus-containing brazing alloys are usually referred to as self-fluxing brazing alloys. When melted in air the products of oxidation form a fluid compound which acts as an effective flux. These alloys are only effective if melted in an oxidising atmosphere. They should not be applied to ferrous or nickel-base alloys since they form brittle compounds which weaken the joint. Typical compositions are shown below.

These alloys are known commercially under various trade names, e.g. *Cupro-Tectic* and *Silbralloy* (Cu-P), and *Silfos* (80:15:5 Cu-Ag-P).

BS 1845 Type	*Ag*	*P*	*Cu*	*Freezing range °C.*
6	13–15	4–6	Balance	625–780
7	—	7·0–7·5	Balance	705–800

3. Brazing Brasses

The oldest and best-known method of brazing involves the use of brazing brasses or 'brazing spelter', using borax as a flux. Typical compositions of brazing brasses are shown in Table 13.2. These alloys melt at higher temperatures than the silver solders but sound joints having tensile strengths of 390–460 N/mm² are possible. Bonding in brazing is due to mutual alloying of the base metal and the brazing alloy.

BS 1845 Type	*Cu*	*Zn*	*Sn*	*Approx. freezing range °C.*
CZ 1	49–51	Balance	—	860–870
CZ 2	53–55	,,	—	870–880
CZ 3	59–61	,,	—	885–890
CZ 4	53–55	,,	0·8–1·2	860–870
CZ 5	59–61	,,	0·8–1·2	880–890

Table 13.2. Brazing Brasses

Brazing of Aluminium

For aluminium brazing special filler alloys of the aluminium-silicon type are used. These are covered by BS 1845.1966.

WELDING

Welding processes may be broadly classified into two types, viz.

1. Fusion-welding processes
2. Solid-phase welding processes

The distinction between these types is summarised in Table 13.3.

Fusion-welding processes	*Solid-phase welding processes*
1. Metals to be joined are locally melted. 2. No pressure applied. 3. Extra metal added in the form of a filler wire or consumable electrode.	1. Metals to be joined do not melt.* 2. Pressure applied to form the joint. 3. No additional metal required.

Table 13.3. Distinction Between Fusion and Solid-phase welding Processes
*In electrical resistance welding some melting takes place.

The various welding processes employed will not be discussed in this book but a classification of the chief processes is shown in Table 13.4.

Fusion-welding processes	*Solid-phase welding processes*
1. Metallic-arc 2. Oxy-acetylene gas 3. Inert-gas shielded arc processes (M.I.G. and T.I.G.) 4. Atomic hydrogen 5. Carbon arc 6. Plasma arc 7. Laser welding 8. Electron-beam	1. Electrical resistance welding (spot, seam, flash-butt welding) 2. Blacksmith's forge welding 3. Explosive welding 4. Cold-pressure welding 5. Friction welding 6. Ultrasonic welding

Table 13.4. Classification of the Chief Welding Processes

Metallurgical Aspects of Welding

The metallurgical aspects of welding are particularly interesting since in a welded joint we have examples of cast, wrought, and heat-treated structures. The weld deposit will possess a typical cast structure with all its inherent defects. The heat-affected zone of the parent metal will exhibit the effects of heat-treatment, whilst the unaffected portion will probably reveal a typical wrought structure. Welded joints may therefore be studied under the following headings.

1. The weld metal deposit
2. The heat-affected zone of the parent metal

1. The Weld Metal

The weld metal is, in effect, a miniature casting which has cooled rapidly from an extremely high temperature. Long columnar crystals may therefore be formed giving rise to a relatively weak structure (Fig. 13.2 (*a*)). In a multi-run weld each deposit 'normalises' the preceding run and considerable grain refinement is obtained with consequent improvement in mechanical properties. In this case only the top run exhibits a coarse 'cast' structure and this can largely be removed after welding if necessary.

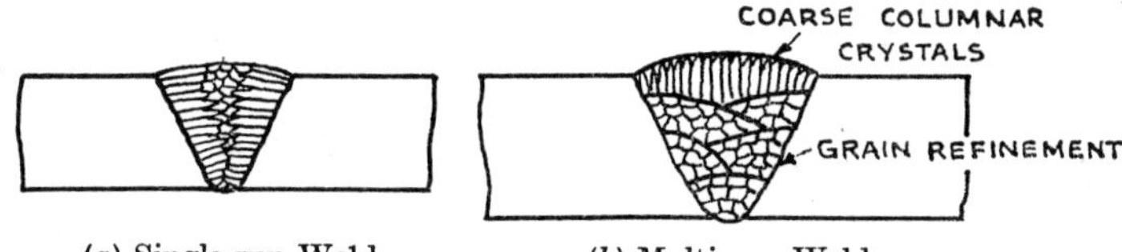

(*a*) Single-run Weld (*b*) Multi-run Weld

Fig. 13.2. Diagrammatic Representation of Structure of Weld Metal in Single and Multi-run Weld Deposit

The effect of the correct welding temperature on the structure of a spot weld is shown in Fig. 13.3.

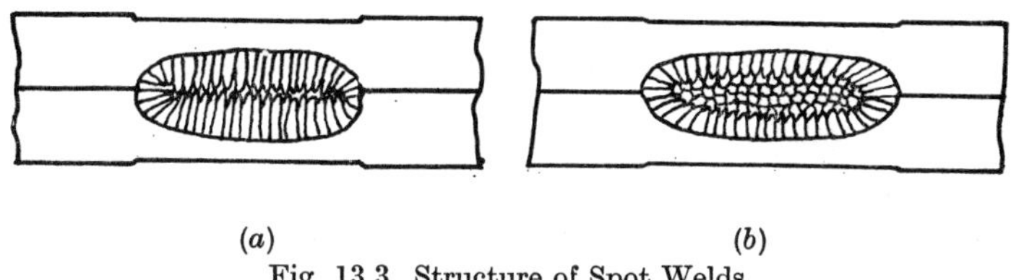

(*a*) (*b*)

Fig. 13.3. Structure of Spot Welds

If the welding temperature is too high the columnar crystals will meet at the centre, forming a plane of weakness (Fig. 13.3 (*a*)) which may lead to intercrystalline cracking. If the temperature is correct then equi-axed grains will form at the centre before the columnar crystals can meet (Fig. 13.3 (*b*)). The importance of correct control of current and time in spot welding is therefore apparent.

Other possible defects in the weld metal include non-metallic inclusions, gas porosity, and cracking.

Non-metallic Inclusions

The formation of oxide and nitride inclusions due to atmospheric contamination is usually avoided by the use of a flux. Modern flux-coated electrodes usually provide good quality weld deposits substantially free from harmful inclusions. In the argon-arc welding process, the metal is deposited under a shroud of the inert gas argon, which prevents oxidation, and no flux is necessary. Slag inclusions can be avoided in multi-run welds by effective removal of the slag after each deposit.

Gas Porosity

The chief cause of gas porosity is the presence of hydrogen in the weld metal, or the reaction of hydrogen with any oxide present in the melted parent metal to form steam. The solubility of hydrogen in most metals such as copper, aluminium and iron varies with temperature as shown in Fig. 13.4.

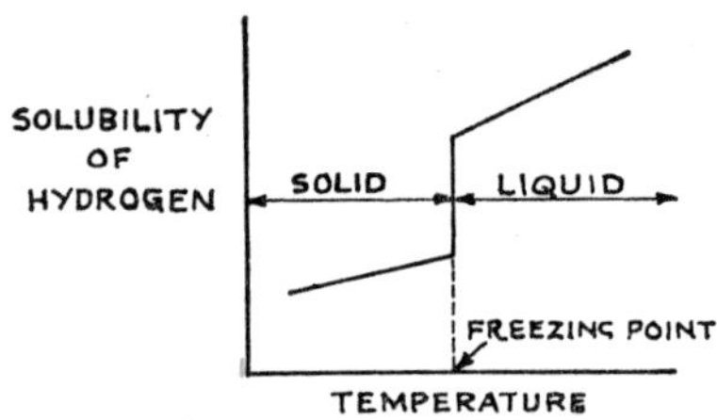

Fig. 13.4. Solubility of Hydrogen in Weld Metal

Hydrogen is readily soluble in the liquid state but only slightly soluble in the solid state. Considerable quantities of hydrogen are therefore evolved upon solidification which may cause gas porosity in the solid weld metal.

In the welding of tough-pitch copper, which contains a small amount of oxygen in the form of cuprous oxide, the hydrogen may react with the oxide of the melted parent metal to form steam. The steam produced by this reaction may give rise to unsoundness.

There are numerous sources of hydrogen in welding, the chief ones being the welding flame in gas welding or the flux coating in metallic arc welding.

Weld-metal Cracking

Welded joints which are prepared under restraint are liable to intercrystalline cracking in the weld deposit due to contractional strains set up during the cooling of the metal. Such cracking, usually known as 'hot cracking', is largely related to the grain size and the presence of grain boundary impurities. At high temperatures the grain boundaries are more able to accommodate shrinkage strains than the grains themselves. A coarse-grained deposit, with large columnar crystals, possesses a relatively small grain boundary area and is therefore more susceptible to hot cracking. The presence of low-melting-point grain boundary films, e.g. ferrous sulphide in mild steel deposits, is also known to give rise to hot cracking. The addition of the correct proportion of manganese results in the formation of manganese sulphide which has a higher melting point and the susceptibility to cracking is thus greatly reduced.

Fundamental work carried out on the welding of aluminium alloys* has shown that weld-metal cracking can be divided into that which occurs above the solidus and that which occurs below the solidus. It was found that cracking of the former type occurred mainly in alloys which solidified over a range of temperature. Pure metals and eutectic alloys were not susceptible to cracking at high temperatures. Sub-solidus cracking is not as frequent as that which occurs above the solidus.

2. The Heat-affected Zone of the Parent Metal

It is not easy to generalise when considering the effect of welding heat on the structure and properties of the parent metal. The extent of any structural change will depend upon the time at temperature, and consequently such factors as the thermal conductivity, specific heat, and dimensions of the plate, together with the speed and method of welding are important. Welded joints prepared from metals of high thermal conductivity, such as copper and aluminium, possess wider heat-affected zones than those prepared from nickel or steel. Metallic-arc welding produces a more concentrated heating effect than gas welding. Increase in the welding speed also reduces the width of the heat-affected zone.

In general with non-ferrous metals and alloys it would be expected that softening of work-hardened or age-hardened metal would occur in the heat-affected zone. The softening effect in age-hardened alloys is usually more sluggish but is a problem in the welding of heat-treated aluminium alloys. In addition grain growth usually occurs in non-ferrous alloys but this is usually insufficient to have any great effect on the strength of the welded joint. The structure of the heat-affected zone of a typical weld preparation in hard-rolled aluminium sheet is shown in Fig. 13.5.

The heat-affected zone in mild steel plate will exhibit various structures, ranging from an overheated structure (Fig. 5.3(*a*)) for those parts heated to well above the upper critical range to an underannealed structure (Fig. 5.3(*b*)) for those parts heated to within

*Aluminium Federation. Research Report No. 2.

the critical range. The welding of mild steel presents no serious difficulties.

In contrast to most non-ferrous alloys an increase in hardness will occur in the heat affected zones of steels. The degree of hardening increases with increasing tensile strength of the steel.

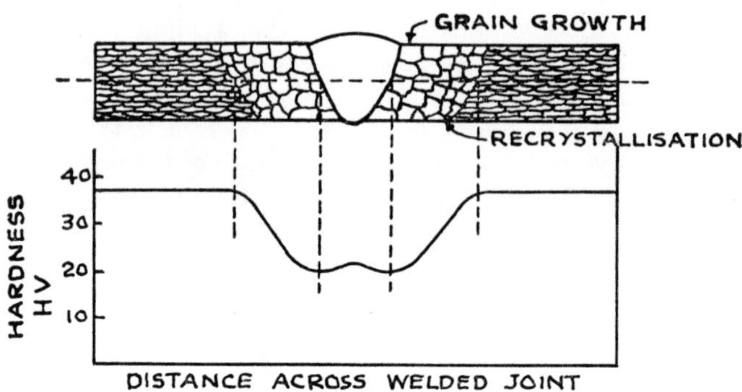

Fig. 13.5. Effect of Welding on the Structure and Hardness of Hard-rolled Aluminium Sheet

Hard-zone Cracking

With certain low-alloy high-tensile steels a hard martensitic structure may form in the heat-affected zone of plate. When the joint is prepared under restraint there is a tendency to cracking in this

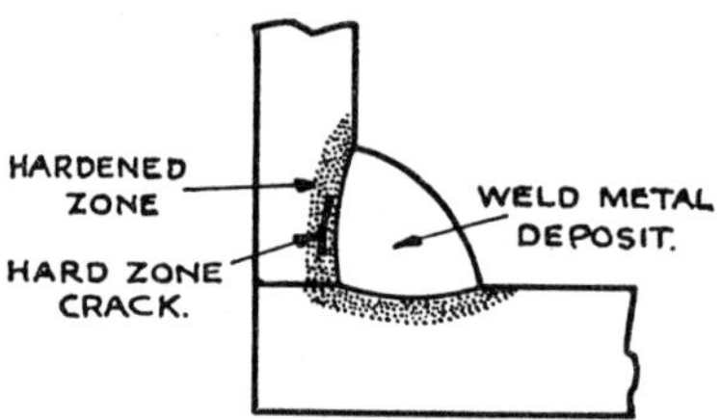

Fig. 13.6. Hard-zone Crack in a Welded Joint in Low-Alloy Steel

hardened zone. The hardness of the heat-affected zone will depend upon the composition of the steel and the rate of cooling. Preheating to about 200°C., the use of large-gauge electrodes, and slow welding speeds, are all effective in reducing hardness and cracking tendency. It has also been found that cracking can be overcome by the use of either low-hydrogen electrodes or fully austenitic electrodes.

The effectiveness of low-hydrogen electrodes suggests that the presence of hydrogen is a factor in hard-zone cracking. Hydrogen is soluble in austenite but insoluble in martensite and ferrite. Hydrogen from the weld deposit diffuses into the heat-affected zone which, during welding, is in the austenitic condition. When transformation to martensite takes place the hydrogen is rejected, collects in microfissures, and builds up a pressure which eventually leads to cracking. The use of austenitic weld metal minimises diffusion of hydrogen since it retains most of the hydrogen in the weld deposit. It is advisable to combine austenitic electrodes with low-hydrogen coatings.

Fortiweld Steel

Because of the danger of hard-zone cracking most of the commercial low-alloy high-tensile steels can only be welded using special procedures. The best combination of mechanical properties and weldability is possessed by a steel known as *Fortiweld.* This steel developed by Appleby-Frodingham Steel Co. Ltd. was first reported in a paper by Bardgett and Reeve in 1949. Fortiweld is a low-carbon molybdenum-boron steel of the following approximate analysis:

C	Mn	Mo	B
0·10–0·16	0·5	0·5	0·0015–0·0035

Fortiweld possesses a yield point of 450 N/mm^2 and a TS of 620 N/mm^2 in the 'as rolled' condition. The good mechanical properties are combined with a weldability almost equivalent to that of mild steel.

Welding of Cast Irons

The welding of cast iron usually involves the repair of fractured castings. Considerable experience is therefore necessary since each repair job usually requires a somewhat different technique.

Cast iron has a low ductility and further cracking is likely to occur due to the stresses set up by the rapid localised heating and cooling. Hard brittle weld deposits may be formed due to absorption of carbon from the molten parent metal. If the cooling rate is rapid the heat affected zone of the parent metal will also be hardened due to the formation of martensite. These defects can be overcome to a large extent by preheating to about 500°–600°C. followed by slow cooling, either in a furnace or under a blanket of sand. In gas welding, the use of a filler wire containing a high silicon content reduces the tendency to form hard white-iron deposits. In metallic-arc welding, if preheating is not possible, a special electrode giving a nickel alloy deposit can be employed. This avoids the formation of a hard deposit since nickel does not form hard compounds with carbon from the parent metal.

Welding of Austenitic Stainless Steels

The problem of carbide precipitation in the heat affected zone of the parent metal during the welding of austenitic stainless steels has already been discussed (page 79). A neutral flame should be employed in oxy-acetylene gas welding, and since a slightly carburising flame could lead to carbon pick-up, this process is only recommended where maximum corrosion resistance is not required. In metallic-arc welding a niobium-stabilised electrode should be employed even if the parent metal has been stabilised with titanium. This is due to the loss of titanium that occurs during the transfer of molten metal across the arc. Argon-arc welding is being increasingly employed since it avoids the use of corrosive fluxes.

Welding of Aluminium and its Alloys

The chief processes employed are the oxy-acetylene gas and argon-arc welding processes. Spot welding is extensively used in the aircraft industry. Among the factors to be considered in the welding of aluminium are the relatively low melting point, high thermal capacity and conductivity, high coefficient of expansion and rapid oxidation of the metal. The use of a flux is essential in gas welding to remove the oxide film, and since flux residues are corrosive they should be removed thoroughly after welding. In gas welding, a strictly neutral flame should be employed. Gas porosity due to hydrogen and tendency to cracking have been discussed previously.

Welding of Copper and its Alloys

Difficulties are encountered in the welding of tough-pitch copper. In the wrought condition, this grade of copper contains oxygen in the form of dispersed particles of cuprous oxide which are not detrimental to the mechanical properties. However, when the copper is remelted, the oxide dissolves, and upon solidification it forms a eutectic at the grain boundaries. This type of structure would be present in the weld metal and in those parts of the parent metal which have been heated to above the solidus temperature. Intermittent welding is employed for tough-pitch copper, with hammering of the hot welds to break up and disperse the oxide. Tough-pitch copper also gives rise to gas porosity in the weld metal due to gas-metal reactions (page 132), and to 'gassing' in the heat-affected zone of the parent metal (page 95). Where possible, deoxidised copper is employed for welding, and the use of a deoxidiser filler wire containing 0·05% P and 1·0% Ag is common practice for all grades of copper. The thermal conductivity of copper is about eight times that of mild steel and preheating is essential in order to obtain fusion.

The chief problem in the welding of brass is the volatilisation of zinc. Zinc boils at 906°C. and consequently vaporises during welding, giving rise to porosity. The vapour oxidises to form a white cloud of zinc oxide, which not only obstructs the vision of the welder, but is also poisonous.

Difficulties are encountered in the welding of aluminium-bronzes due to the formation of the tenacious film of aluminium oxide. However, satisfactory arc welding of these alloys has been made possible by the development of suitable electrodes.

Welding of Nickel and its Alloys

The physical properties of nickel such as melting point, thermal capacity, conductivity and thermal expansion and contraction are very similar to those of steel. Welding of nickel alloys presents few difficulties provided the weld is prepared without undue restraint. Nickel is particularly sensitive to sulphur attack (page 117) and thorough cleaning of the surface prior to welding is important.

14. The Corrosion of Metals

Because of its costly destructive effect the study of corrosion, and its prevention, is of vital importance to both the metallurgist and the engineer. Corrosion is a complex problem due to the many variables involved. The factors governing the rate of corrosion may be broadly divided into those relating to (1) the metal and (2) the environment.

The factors relating to the metal or alloy include (*a*) the position of the metal in the electro-chemical series; (*b*) contact with dissimilar metals; (*c*) microstructure, e.g. the presence of impurities or of a second constituent; and (*d*) the presence of internal stress. The factors involved when considering the effect of environment include (*a*) relative humidity; (*b*) presence of impurities in the atmosphere; (*c*) rate of supply and distribution of oxygen; (*d*) rate of flow of liquid; (*e*) the acidity or alkalinity of the liquid; and (*f*) presence of external stress. These are only a few of the factors involved but are sufficient to emphasise the complex nature of the problem.

In general two principal forms of corrosion can be distinguished:

1. Direct chemical corrosion
2. Electro-chemical corrosion

1. DIRECT CHEMICAL CORROSION

This type of corrosion usually involves direct combination between the metal and dry gases such as oxygen, sulphur dioxide, and chlorine, and usually occurs at high temperatures. The nature of the film produced has an important effect on the extent of subsequent corrosion. If the film is hard, adherent and protective, then corrosion will eventually cease, as in the case of the corrosion and heat-resisting alloys of the nickel-chromium type.

2. ELECTRO-CHEMICAL CORROSION

This type of corrosion covers all forms of 'wet' corrosion, i.e. where the metal is in contact with a liquid or even a moist atmosphere. In the electro-chemical theory it is assumed that all metals have a tendency to dissolve or corrode, when the metal discharges positively-charged particles, called ions, into solution. This leaves the metal with a characteristic negative charge or potential. The greater the negative potential the greater is the tendency of the metal to dissolve or corrode. The corrosion resistance of metals is governed by their position in the electro-chemical series in which metals are arranged according to their electrode potentials (Table 14.1). The value of these potentials is very small and their results are expressed relative to hydrogen which is taken as zero.

Anodic end (corroded)	
Metal	*Electrode potential volts*
Sodium	−2·71
Magnesium	−2·40
Aluminium	−1·70
Zinc	−0·76
Chromium	−0·56
Iron	−0·44
Cadmium	−0·40
Nickel	−0·23
Tin	−0·14
Lead	−0·12
Hydrogen	0·00
Copper	+0·35
Silver	+0·80
Platinum	+1·20
Gold	+1·50
Cathodic End (protected)	

Table 14.1. The Electro-chemical Series of Metals

The tendency of each individual metal to corrode is relatively small, but is greatly increased when it is in contact with a dissimilar metal in the presence of a conducting liquid, referred to as the electrolyte. A current will flow between the two metals since they are at different potentials. Corrosion of the one higher in the table (known as the ANODE) will be accelerated whilst the metal lower in the table (known as the CATHODE) will be protected. The rate of corrosion is governed by the relative areas of the anode and the cathode. In general for a given area of anode, the attack increases in severity the greater the area of the adjacent cathode.

It will be seen that for electro-chemical corrosion to occur there must be a cathode, an anode and an electrolyte. The formation of an anode and cathode need not necessarily be due to contact with a dissimilar metal. Electrolytic corrosion can also occur due to the presence of an impurity or second constituent in the structure of the alloy, or to a local difference in oxygen concentration on the surface (differential aeration effect).

Relationship between Microstructure and Corrosion-resistance

For maximum corrosion-resistance an alloy should be of the single solid-solution type. The presence of a second constituent or impurity can reduce the corrosion resistance due to electrolytic action. 'Weld decay' in austenitic stainless steels (page 78) is an example of the decrease in corrosion-resistance due to the precipitation of chromium-rich carbides from the solid-solution austenite. The corrosion resistance of some commercially pure metals such as magnesium can be greatly increased by decreasing the amount of impurities such as iron. When present in amounts greater than 0·003% the iron is present as a second constituent and this gives rise to electrolytic corrosion.

Differential Aeration Effect

The work of Dr. U. R. Evans of Cambridge has shown that corrosion can take place even in a pure metal if there is a difference in oxygen concentration on the metal surface. This can be demonstrated using the apparatus shown in Fig. 14.1. If two pieces of the same metal are immersed in an electrolyte, no current is detected since there is no difference in potential between them. When one piece of metal is aerated a current is detected, the aerated metal being protected (cathode) whilst the non-aerated metal corrodes (anode). When the current of oxygen is diverted to the other compartment the direction of the current is reversed. This theory can be used to explain corrosion problems such as 'pitting', and also why metals corrode under scale deposits or along cracks and scratches more readily than elsewhere on the surface. A differential aeration cell could be set up when metal sheets are stacked together with water trapped between them. This could be due to differences in oxygen content among the pockets of liquid held between the sheets by surface tension.

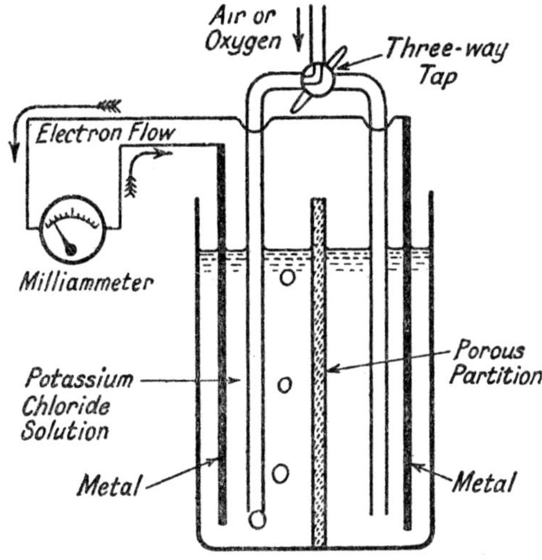

Fig. 14.1. Experiment to illustrate Differential Aeration Effect. From *Introduction to Metallic Corrosion* by U. R. Evans (Edward Arnold)

Pitting

Pitting is an example of the differential aeration effect. The initial depression or pit in the surface may be the result of several factors, e.g. a break in a protective film or scale, or the solution of a non-metallic inclusion due to electrolytic action. Once a pit is formed the corrosion proceeds rapidly since the surface of the metal (cathode) has a greater access to oxygen than the base of the pit (anode) (Fig. 14.2). Corrosion is accelerated by the fact that the surface area of the cathode is considerably greater than that of the anode.

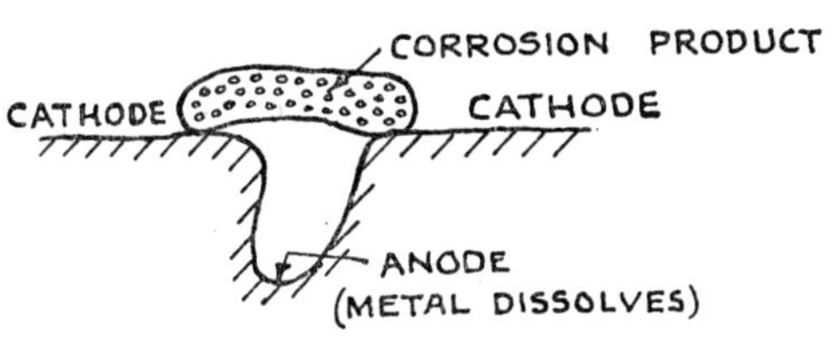

Fig. 14.2. 'Pitting' Corrosion

The corrosion product accumulates at the mouth of the pit and assists corrosion by making oxygen diffusion more difficult.

The Influence of Stress in Corrosion

The influence of stress in corrosion can be seen in such corrosion phenomena as stress corrosion, cavitation erosion, fretting corrosion and corrosion fatigue. Corrosion fatigue will be discussed later (page 184) under the heading of fatigue.

Stress Corrosion

This is due to the joint action of sustained static stress and corrosion. The stresses involved are usually internal stresses produced by some previous treatment such as cold working. The failure may be transcrystalline or intercrystalline but is predominantly one or the other. The conditions for stress corrosion are highly specific and the rate of crack propagation is much greater than the average rate of the normal corrosion processes. Stress corrosion cracking may be avoided by stress relief annealing prior to service.

One of the earliest and best known examples is that of the 'season cracking' of brass cartridge cases. Season cracking is the name given to the intercrystalline cracking that occurs when work-hardened brasses are exposed to mildly corrosive media, e.g. industrial atmospheres. It can be avoided by annealing at about 250°C which relieves the internal stresses without softening the alloy.

The commonest environments producing stress-corrosion cracking for various alloys are as follows:

Alloy	*Environment*
Brass	Ammonia
Mild steel	Caustic alkalies and nitrate solutions
18/8 Stainless steel	Chloride and caustic solutions. Steam
Aluminium alloys	Marine atmospheres

Cavitation Erosion

Cavitation erosion may occur with materials in rapidly moving liquid environments, e.g. marine propellors and rudders, and in pumps and pipe-work circulating water. During the flow divergence of the liquid millions of cavities are formed in regions of low pressure. The subsequent rapid collapse of these cavities produces a strong shock wave which damages the adjacent metal. During static periods corrosion of the freshly exposed damaged surfaces will also occur and accentuate the damage.

With closed systems such as diesel cooling circuits it is advisable to use inhibitors in the liquid. In other cases the use of materials which combine good wear and corrosion resistance is essential. Such materials include 18/8 Cr/Ni stainless steel, which can be used as a welded overlay, 10% aluminium-bronze and stellite.

Fretting Corrosion

Fretting corrosion occurs frequently in vibrating machinery where two parts in contact are subject to small repeated movements relative to each other. Severe pitting of the surfaces occurs and fine oxide particles are produced by abrasive wear. In steel these particles are reddish-brown in colour and are often referred to as 'cocoa'. In aluminium and magnesium alloys the wear particles are black. The result of fretting is a serious loss of dimensional accuracy and a reduction in the fatigue strength.

When the two surfaces are in contact adhesion will occur at certain points. When the vibration takes place fracture will occur away from these cold welds producing metallic wear particles. These particles subsequently oxidise thereby increasing in volume and thus accentuating the abrasive wear. Fretting damage will increase with pressure until the pressure is sufficiently high to prevent slip when fretting is avoided altogether.

If the surfaces are not intended to move relative to each other then fretting damage can be avoided by preventing slip. This can be done by removing the source of vibration, increasing the pressure or by introducing an elastic material such as rubber between the surfaces. However, if slip cannot be prevented then it is preferable to use treatments which reduce the friction. These treatments include the use of :

(i) solid lubricants, such as a bonded layer of molybdenum disulphide
(ii) plastic coatings such as P.T.F.E.
(iii) impregnated anodised coatings for light alloys
(iv) phosphated coatings impregnated with a lubricant for steels

All materials are susceptible to fretting and fretting corrosion frequently occurs in shrink fits, press fits and bolted assemblies, splined couplings, keyed gears and the seatings and tracks of ball and roller races.

Passivity

Certain alloys, particularly stainless steels, are capable of forming a thin protective oxide film when used in contact with an oxidising medium such as nitric acid. The alloy is said to become 'passive' since corrosion is greatly minimised. This film does not offer such good protection against corrosion by less oxidising acids such as hydrochloric or sulphuric acid.

Effect of Design on Corrosion

The importance of correct design cannot be over-emphasised. Contact with the corrosive medium should be reduced to a minimum. Adequate drainage and ventilation should be provided, and the design should aim to prevent crevices or moisture traps. Attention should be paid to sealing of joints since if aqueous liquid cannot intrude there can be no corrosion. In this connection butt welded joints are generally better than riveted joints.

PROTECTION OF METALS AND ALLOYS FROM CORROSION

Certain alloys possess an inherent resistance to various types of corrosion, e.g. stainless steel, nickel, Monel and Inconel. However, these alloys are expensive and their use cannot be economically justified except for special applications such as chemical plant construction. It is therefore necessary to adopt various methods to protect metals and alloys from corrosion. The methods used may be broadly classified as follows:

1. Metallic Coatings

These coatings may be applied by (*a*) dipping, (*b*) electro-deposition, (*c*) cladding, (*d*) spraying, (*e*) cementation methods such as calorising, sherardising and chromising.

2. Non-metallic Coatings

These include oxide and phosphate films and protection by various paints, varnishes and lacquers.

3. Cathodic Protection

The metal is sacrificially protected by contact with another metal which is higher in the electro-chemical series.

Each of these methods of protection will now be discussed in more detail.

1. Metallic Coatings

(a) Hot Dipping

This method is used in the manufacture of tin plate and of galvanised-iron sheet. The thoroughly cleaned steel is passed through a layer of molten flux into a bath of molten tin or zinc. A thin layer of tin or zinc adheres to the steel due to the formation of a thin zone of an iron-tin or iron-zinc alloy. The process of tin-plating by dipping has to a large extent been replaced by electrolytic tinning. Tin-plate with a coating of the order of 0·001 25 mm thick is used extensively in the canning industry.

Zinc is anodic to steel and therefore if the coating is broken the zinc will corrode sacrificially. This is an example of the sacrificial protection of steel. On the other hand tin is cathodic to steel and the protection is purely mechanical. The tin coating must be of good quality since if it is ruptured then corrosion of the steel will be accelerated by electrolytic action. The differences in the two types of protection is illustrated in Fig. 14.3.

(b) Electro-deposition or Electro-plating

The component to be plated is made the cathode in an electrolytic cell. The plating solution, or electrolyte, consists usually of a salt of the plating metal together with various special additions. The anode consists of the plating metal, which, when the current is passed, dissolves and plates out on the cathode. In some cases, as in chromium plating, an insoluble anode may be used, when the chromium is provided by the electrolyte itself.

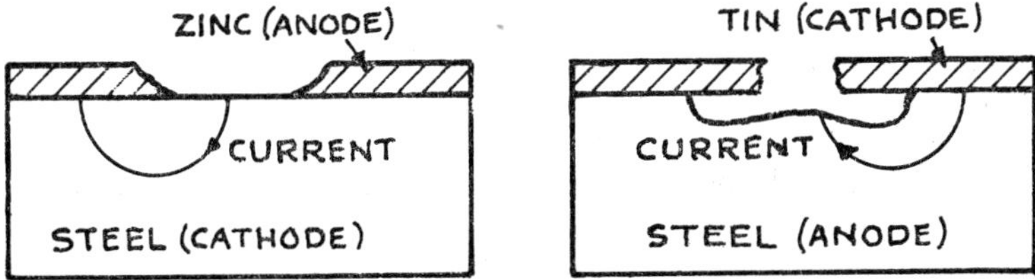

Fig. 14.3. Diagrams to Illustrate the Difference in the Type of Protection given to Steel by coatings of Zinc and Tin

The usual thickness of deposits of the commonly plated metals are nickel (0·007 5 mm–0·050 mm), chromium (0·000 25 mm–0·001 25 mm), copper (0·007 5 mm–0·025 mm), cadmium (0·005 mm–0·012 5 mm), and silver (0·007 5 mm–0·030 mm).

No alloying is involved and adhesion depends entirely upon the intimate contact of the coating and base metal. The components for plating must therefore be thoroughly cleaned.

(c) Cladding

The use of clad steels was developed for the chemical industry in order to avoid the expense of using thicker plates of corrosion-resistant materials such as nickel, Monel, and stainless steel. Clad steels are manufactured by hot-rolling composite billets, to produce a cladding thickness equal to 5–20% of the total thickness of the plate. Superficial alloying takes place at the interface.

Aluminium alloys are frequently protected from corrosion by a cladding of pure aluminium, applied by hot-rolling. The usual clad thickness is about 10% of the total thickness.

(d) Spraying

Metallic coatings of aluminium, zinc, tin, copper, lead, brass and bronze may be applied by spraying. The molten metal is sprayed from a pistol in which the metal in wire form is melted by an electric arc or oxy-acetylene flame and blown out by compressed air. The adherence of sprayed coatings to the base metal is lower than the other methods previously mentioned. No alloying occurs and the surface should be clean and preferably roughened. The usual thickness of zinc and aluminium coatings is 0·1 mm–0·3 mm.

(e) Cementation Processes

Examples of cementation processes are sherardising, calorising and chromising. In each case the component is surrounded by powdered metal and heated when alloying of the two metals occurs.

Sherardising. A uniform coating of zinc is obtained by heating the component in zinc dust at 350°–375°C. for three to twelve hours. After three hours a film thickness of approximately 0·063 mm is obtained.

Calorising. A layer of an iron-aluminium alloy approximately 0·625–0·75 mm in thickness is obtained by heating mild steel in powdered aluminium at 850°–1 000°C. Calorised steel has a good resistance to oxidation at elevated temperatures.

Chromising. A chromium-rich surface is obtained by heating the steel in a powdered mixture of aluminium oxide and chromium in an atmosphere of hydrogen at 1 300°–1 400°C. for three to four hours. The hydrogen is necessary to prevent oxidation of the chromium.

2. Non-metallic Coatings

(a) Anodising of Aluminium Alloys

Aluminium possesses a good resistance to corrosion due to the formation of a hard self-healing adherent oxide film on the surface. The object of anodising is to increase the thickness of this film. This is achieved by making the component the anode in an electrolyte of either chromic, sulphuric or oxalic acid. The cathode consists of either stainless steel or lead. The oxide film varies in thickness according to the process used, but is normally 0·000 25–0·005 mm thick. The sulphuric acid process produces the thicker film and takes about 20–40 minutes. The chromic acid process produces a thinner, harder film and takes about one hour.

The anodised skin can be coloured by immersion in a solution containing a suitable dye. Sealing of the anodised skin is essential after anodising in order to improve the corrosion-resistance and hardness. This can be achieved by immersion in boiling water for about 20 minutes or by the application of lanolin.

Hard anodic films approximately 0·05 mm in thickness are used for wear resistance. In order to build up such thick films a low temperature is necessary to keep the resistance of the bath low. This is usually achieved by refrigeration.

(b) Chromating

Oxide films can be produced on magnesium alloys by dipping in solutions of potassium dichromate together with other additions. Chromate surfaces vary in colour but are usually grey or black. The treatment is generally followed by painting with a zinc chromate paint.

(c) Phosphating

Phosphating treatments, such as *Parkerising*, *Bonderising*, *Granodising* and *Walterising* are usually carried out on steels and zinc-base alloys. A coating of phosphate is produced by immersion in a solution of acid iron, manganese, or zinc phosphates together with various accelerators. The time of immersion varies from 1–10 minutes. The surface produced is treated with varnish, paint or lacquer to improve the corrosion resistance. Phosphating treatment is carried out on motor car, cycle and refrigerator parts prior to enamelling.

3. Cathodic Protection

Metals such as magnesium are frequently employed as galvanic anodes for the protection of steel. Magnesium is higher than iron in the electro-chemical series and corrodes sacrificially. Cathodic protection is given to pipe lines, buried structures, and the hulls and compartments of ships.

Cathodic protection of pipe lines is often carried out using the impressed voltage method. In this method a small d.c. voltage applied to the pipe will provide sufficient electrons to make the pipe a cathode and so prevent corrosion.

15. The Metallographic Examination and Non-destructive Testing of Metals and Alloys

One of the functions of a metallurgist attached to an engineering works is the examination of defects and service failures in metal components. In addition he is required to carry out various routine tests on both the incoming material and the manufactured product. It is therefore important that the engineering student should have some knowledge of the various methods of examination and testing, particularly their scope of application and limitations. The various methods of mechanical testing are included in the next chapter.

THE EXAMINATION OF FRACTURES

Valuable information regarding the cause of failure can often be obtained by a visual examination of the fracture. The presence of slag inclusions, porosity and blow holes is readily observed in the fractures of welded joints and castings. Fractures occurring during the hot-working of ingots are frequently the result of coarse columnar crystals forming planes of weakness (Fig. 1.11) due to the casting temperature being too high. Components subjected to alternating or reversed stresses are liable to fatigue failure. A fatigue fracture is usually characterised by two well-defined zones as shown in Fig. 15.1. The smooth discoloured portion is evidence of the gradual extension across the section of a crack, which frequently starts from some point of stress-concentration, e.g. a sharp fillet, tool mark, oil hole or slag inclusion. Once a crack has been initiated the applied load becomes more concentrated, causing gradual extension of the fracture. Finally the remaining area cannot sustain the load, and breaks suddenly, giving a fibrous fracture. The difference between these two zones is not so apparent in the more brittle materials.

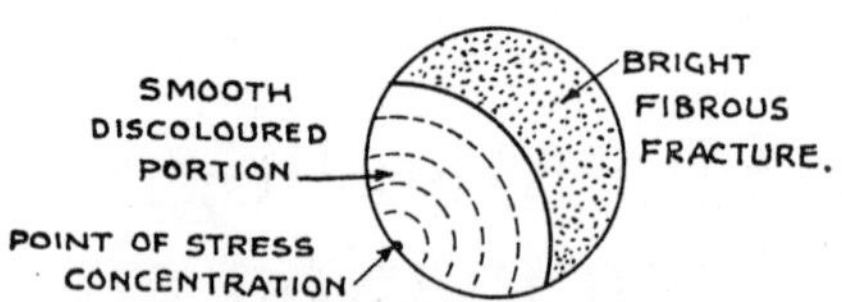

Fig. 15.1. Typical Fatigue Fracture

MACRO-EXAMINATION

Macro-examination involves the examination of prepared sections either with the naked eye or using a small hand magnifying lens. The surface to be examined need only be ground to about 0 or 00 emery paper and washed to remove the grit, before being deeply etched, using the appropriate etching reagents.

This method is useful in revealing flow lines in forgings (Fig. 2.11), grain size in castings, the contour of weld deposits and heat-affected zones in welded joints. Gross defects such as slag, porosity and blow holes in welds and castings are also readily visible without resorting to microscopical examination.

MACRO-ETCHING REAGENTS

Steel

(1) 50% Hydrochloric Acid

The specimen should be immersed in the boiling reagent for 10–45 minutes. In general the higher the carbon content the longer the etching times required. The reagent is used to reveal flow lines, the structure of welds, cracks and porosity.

(2) 25% Nitric Acid

This reagent is similar to (1) but is used cold for large surfaces which cannot be conveniently heated.

(3) 10% Ammonium Persulphate Solution

This reagent should be made up fresh and applied as a cold swab with absorbent cotton wool. It is useful in revealing grain growth in the examination of welded joints.

Copper and its Alloys

(1) Acid Ferric Chloride Solution

25 grams ferric chloride, 25 ml. hydrochloric acid, 100 ml. water.

Used for revealing the dendritic structure of the α solid solution.

(2) Ammoniacal Ammonium Persulphate Solution

Ammonium hydroxide (0·880) 1 part, ammonium persulphate (2·5% solution) 2 parts, water 1 part.

Used for alloys containing the β constituent.

Aluminium and its Alloys

20% Hydrofluoric Acid Solution

The specimen should be degreased with carbon tetrafluoride and then immersed in hot water, prior to swabbing the surface with the etching reagent.

N.B. Particular care should be taken in handling hydrofluoric acid. The acid or its fumes should not be allowed to make contact with the skin or eyes. Etching reagents which contain concentrated acids should only be made up by a responsible laboratory steward or teacher.

SULPHUR PRINTING

The distribution of sulphide inclusions in steels is usually determined by means of a sulphur print. The underlying principle of the test is that dilute sulphuric acid attacks the manganese sulphide inclusions (and ferrous sulphide, if present), liberating hydrogen sulphide gas. This gas reacts with the silver bromide photographic paper, forming a brown stain of silver sulphide. The reactions involved may be represented by the following equations:

$$MnS + H_2SO_4 = MnSO_4 + H_2S$$
$$2AgBr + H_2S = Ag_2S + 2HBr$$

It is not possible to obtain sulphur prints from high alloy steels or non-ferrous metals.

The test is carried out by grinding the surface to be examined to about No. 0 grade of emery. It is then thoroughly cleaned and degreased. A piece of bromide photographic paper is soaked in a 3% solution of sulphuric acid in water for about three minutes. This time should not be exceeded since the gelatine tends to swell and the paper becomes very slippery and difficult to keep in position when applied to the steel. The paper is removed from the acid and then lightly pressed between two sheets of blotting paper to remove excess acid. The emulsion side of the paper is then placed in contact with the surface to be examined. With large specimens the paper is placed face downwards on the specimen and usually rolled with a squeegee to obtain close contact and remove air bubbles. With small specimens it is more convenient to place the paper on a flat surface and to press the specimen firmly against it. The time of contact varies with the steel but is usually about two minutes. The intensity of the brown stain can be judged by lifting the corner of the paper. The print is then rinsed in water and 'fixed' in a solution of hypo for ten minutes. It is then washed in running water for twenty minutes and dried.

It is sometimes possible to obtain a second or even a third print from the same surface without further preparation other than washing and drying. The time of contact required, however, is considerably longer.

THE MICROSCOPIC EXAMINATION OF METALS AND ALLOYS

We have seen that useful information can be obtained about the structure of the metal by macro-examination. However, in order to obtain more detailed information, such examination should be supplemented by microscopical examination. Since microscopical examination of metals depends upon the reflection of light from a polished surface, the preparation of the specimen is very important.

Preparation of Specimens

A representative sample should first be obtained from the material being examined. This should be a convenient size for polishing, about 12–25 mm square or diameter where possible. The difficulties encountered in polishing very small specimens may be overcome by mounting them in plastic or a fusible alloy, e.g. Wood's alloy.

Grinding

The surface to be examined usually contains saw marks and is not perfectly flat. These saw marks can be removed and the surface levelled using a fairly coarse file, or where available a motor-driven linisher. Care should be taken not to overheat the specimen. When the saw marks have been removed the specimen is washed, dried and subjected to fine grinding on a series of emery papers of increasing fineness, i.e. 2, 1, 0, 00, 000 grades of emery. This may be carried out by placing the emery paper on a flat glass plate and holding the specimen so that the scratches are formed in one direction only. Grinding is continued until the scratches left from the previous grinding operation are removed. The specimen is then swilled with water and dried before being transferred to a finer paper. The grinding on the next emery paper is carried out so that the scratches are at right angles to those of the previous paper. Finally a series of fine scratches is obtained in the 000 grade of emery paper, and after washing and drying the specimen is ready for polishing.

For soft metals, e.g. aluminium and magnesium, it is necessary to treat the emery paper with some form of lubricant to prevent particles of emery being embedded in the surface. Typical lubricants are paraffin or a solution of white wax in benzene. It is also advisable to use emery papers that are well worn and smooth for the softer metals and to apply a very light pressure. Suitable wet pregrinders are available on the market which contain strips of emery paper of varying grades clamped to a glass plate. These papers are continually fed with running water which washes away coarse particles. Where large numbers of specimens have to be prepared emery paper fitted to a circular rotating disc may be used.

Polishing

The fine scratches from the 000 grade of emery paper can be removed by using polishing powders on selvyt cloths. The selvyt cloth may be stretched out on a piece of plate glass or attached to the rotary disc of a polishing machine. Various polishing powders are employed, e.g. jeweller's rouge, alumina, magnesia, and chromic oxide. The use of diamond dust in the form of proprietary diamond pastes with special lubricants produces particularly good results. Metal polishes such as Silvo and Brasso may also be used for copper, brass and soft alloys. After polishing, the specimen should be swilled with water, followed by methylated or industrial spirits and finally dried. Drying can be conveniently carried out using a domestic hot-air hair dryer.

Examination of the Polished Surface

The polished specimen, with its 'mirror' finish, free from scratches, should then be examined under the microscope in order to observe any of the following:

(1) Cracks
(2) Blow holes
(3) Hard constituents which stand out in relief, e.g. cementite in white cast iron
(4) Non-metallic inclusions, e.g. slag in wrought iron, graphite in cast iron, or manganese sulphide in steels.

At this stage the microstructure of the metal or alloy is not apparent. Light rays from the microscope strike the surface normally and so retrace their path back eventually to the eye.

Etching

In order to examine the microstructure of the metal or alloy the specimen must be etched. Etching involves the selective corrosion of the polished surface, which renders the various constituents visible by a contrast effect. The specimen is immersed in a suitable etching solution for a given time, removed using nickel or stainless steel tongs, swilled with water and meths, and finally dried. It is then ready for microscopical examination.

Certain corrosion-resisting alloys are frequently etched by means of electrolytic attack. This involves passing a current through an electrolyte containing the specimen as anode and an inert cathode of platinum or graphite.

ETCHING REAGENTS FOR MICROSCOPICAL EXAMINATION

Plain Carbon Steels and Cast Irons

1. Nital

2–5% nitric acid in ethyl or methyl alcohol. The etching time varies from a few seconds up to one minute. This reagent is effective in revealing the ferrite grain boundaries but tends to over-etch the pearlite in so doing.

2. Picral

4% picric acid in ethyl alcohol. This reagent does not reveal the grain boundaries but is more effective in revealing pearlite and spheroidised structures.

Alloy Steels

1. Mixed Acids in Glycerol

Nitric Acid 10 ml
Hydrochloric Acid 20 ml
Glycerol 20 ml
Hydrogen Peroxide 10 ml

Suitable for nickel-chromium alloys, austenitic stainless steels and high-speed tool steels. Specimen should be warmed in hot water before immersion.

2. 10% Oxalic Acid Solution (Electrolytic Etch)

The specimen is made the anode with a cathode of stainless steel, graphite or platinum. Use 6 volts for 10–40 seconds. Used to reveal the grain boundaries of 18/8 stainless steels.

Copper and its Alloys

1. 10% Ammonium Persulphate Solution

Must be freshly prepared. Useful for brasses, bronzes and nickel silvers.

2. Acid Ferric Chloride Solution

Ferric Chloride 10 gm
Hydrochloric Acid 30 ml
Water 200 ml

Suitable for $\alpha\beta$ brasses, bronzes, aluminium-bronzes and cupro-nickel alloys. Darkens the β phase and thus gives more contrast.

Aluminium and its Alloys

1. 1% Sodium Hydroxide Solution

The solution is swabbed over the specimen for 10 seconds.

2. 0·5% Hydrofluoric Acid Solution

This reagent is applied by swabbing with absorbent cotton wool.

3. Keller's Etch

Hydrofluoric Acid 1 ml
Hydrochloric Acid 1·5 ml
Nitric Acid 2·5 ml
Water 95 ml

This reagent is used for duralumin type alloys. The specimen is usually immersed in the reagent for 10–20 seconds and finally washed in warm water.

Nickel and Monel

The following solution should be freshly prepared before use.

Nitric Acid 50 ml
Glacial Acetic Acid 50 ml

Tin and its Alloys

1. Nitric Acid 1 part
Acetic Acid 1 part
Glycerol 8 parts
Use at 40°C. Suitable for tin-lead alloys.

2. Nitric Acid 1 part
Acetic Acid 3 parts
Glycerol 5 parts
Use at 40°C. Suitable for pure tin.

3. Acid Ferric Chloride Solution

Ferric Chloride 10 gm
Hydrochloric Acid 2 ml
Water 95 ml

Used for tin-base bearing metals. Immersion time up to 5 minutes at room temperature.

Lead and its Alloys

1. Glacial Acetic Acid 3 parts
   Nitric Acid 4 parts
   Water 16 parts

Use at 40°C. Etching time depends upon the depth of distorted metal on the prepared surface and varies from 4–30 minutes.

2. Glacial Acetic Acid 3 parts
   30% Hydrogen Peroxide Solution 1 part

Used for lead-antimony alloys. Etching time 5–20 seconds.

NON-DESTRUCTIVE TESTING OF METALS

The development of methods of non-destructive testing has increased considerably over the past twenty years. Such methods enable an assessment to be made of the quality of the finished component without the use of a test-piece or damaging the component in any way.

Non-destructive tests may be roughly classified as follows:

1. Detection of Cracks at or Near the Surface

(*a*) Penetration Methods
(*b*) Magnetic Crack Detection
(*c*) Electrical Methods

2. Detection of Submerged Defects

(*a*) Radiography (i) X-ray Examination
(ii) γ-ray Examination
(*b*) Acoustical Methods (i) Sonic
(ii) Ultrasonic

1(a) Penetration Methods

Penetration methods involve the introduction of a liquid into the surface cracks by immersion in a suitable penetrant. The penetrant enters the crack and the surface is thoroughly cleaned and dried. The position of the crack is betrayed by subsequent seepage of the penetrant.

Various techniques are employed to obtain good contrast between the stain produced and the background. In the chalk test, the component is soaked in a penetrant such as paraffin, and, after cleaning and drying, the surface is dusted with fine dry chalk. Subsequent seepage of the paraffin from the cracks stains the chalk. Improved contrast can be obtained by the addition of suitable dyes, usually red, to the penetrant. Fluorescent penetrants are also available which indicate the presence of cracks by fluorescence when examined under ultra-violet light.

Surface cracks are also revealed after the anodising of aluminium alloys (page 147) using chromic acid as the electrolyte. Seepage of the chromic acid from the cracks gives rise to yellow-brown stains on the surface.

Penetrant methods are not as sensitive as magnetic crack detection, but can of course be applied to both magnetic and non-magnetic materials.

1(b) Magnetic Crack Detection

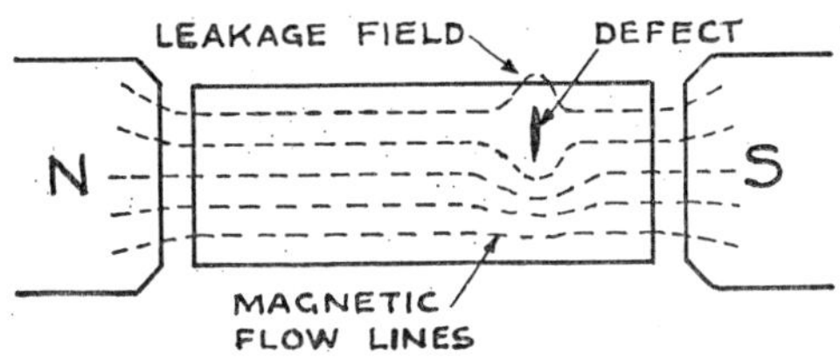

Fig. 15.2. Magnetic Crack Detection

Magnetic crack detection is based on the principle that the defect or crack has a lower magnetic permeability than the metal itself. The magnetic lines of force are thus distorted by the presence of the defect causing a local increase in the magnetic field, known as a leakage field. This leakage field is sufficiently strong to attract magnetic particles, thus revealing the presence of the defect.

In practice the magnetised component is generally immersed in a liquid containing finely powdered magnetic iron oxide in suspension.

The position of the magnetic field relative to the defect is very important (Fig. 15.3). The best indications are obtained when the magnetic field is at right angles to the crack.

Magnetic crack detection is widely used for detecting cracks in steels, but obviously cannot be employed for non-magnetic alloys. Sub-surface defects may be indicated but the sensitivity falls off rapidly with the depth of defect.

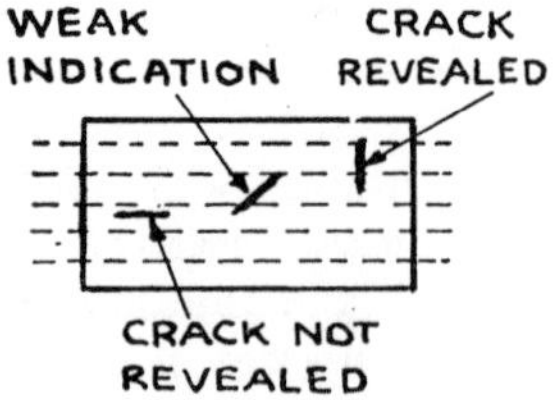

Fig. 15.3. Effect of Position of Magnetic Field relative to the Crack

1(c) Electrical Methods

Electrical methods detect cracks by the observation of their effect on the distribution of an electric current flowing in the metal. Two methods may be distinguished:

(i) Direct current resistance method
(ii) Eddy-current method.

In the former method the potential difference between two electrode probes which make contact with the surface is measured using a sensitive spot galvanometer. The presence of a crack between the electrodes causes an increase in potential difference which is revealed by the galvanometer. This method is limited to clean, smooth surfaces since good contact between the electrodes and the surface is essential.

In the eddy-current method a search coil, carrying alternating current, is moved over the surface of the metal and any change in the impedance of the coil is assumed to indicate the presence of cracks. This method is limited to uniform components in the form of bar, strip or tube, particularly for non-magnetic materials. The interpretation of results is difficult and requires considerable experience.

2(a) Radiography

(i) *X-ray Examination*

This method consists of passing a beam of X-rays through the metal component and studying the effect produced on a photographic film. Defects, such as gas holes and slag inclusions, offer less resistance to the X-rays than sound metal, and their position is therefore indicated on the film by a shadow effect.

X-ray examination is suitable for the detection of gross defects of dimensions of approximately 1% of the total thickness when the greatest length of the defect lies in the beam direction. Unless the orientation of the defect is known by experience it may be necessary to take the radiographs in a number of directions. X-ray radiography is used extensively for the examination of welded pressure vessels and for gas-turbine blades and aluminium castings. Defects such as gas holes, porosity, slag inclusions, shrinkage cavities and large cracks can be readily detected, but it is not considered a satisfactory method for the detection of very fine cracks. For these, ultrasonic testing is a more reliable method. X-ray examination is expensive and is only used where the cost is justified. Considerable experience is needed for the satisfactory interpretation of a radiograph.

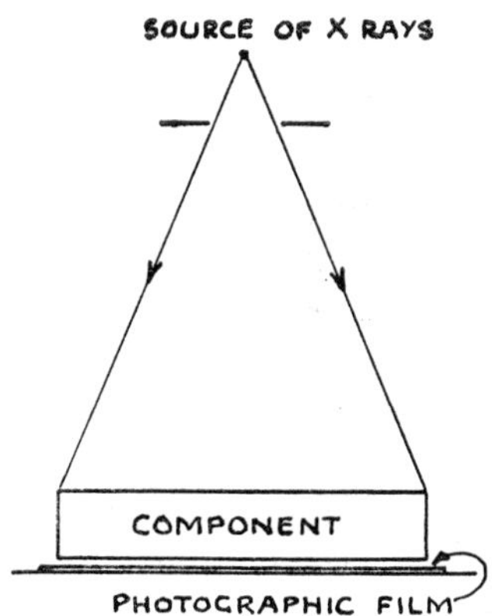

Fig. 15.4. X-ray Examination of Metals

(ii) *γ-ray Examination*

Radiographic examination using γ rays may also be employed, and is capable of detecting the same type of defects as that revealed by X-rays. The method is usually slightly less sensitive, the sensitivity ranging from 1–2%. γ rays are generated continuously by radioactive materials and unlike X-rays, the γ-ray source cannot be 'turned off' after use. Care must be taken to shield personnel from the effects of γ rays.

2(b) Acoustical Methods

(i) *Sonic Methods (e.g. The Hammer Test)*

Gross internal defects increase the damping capacity of a metal component. When the metal is hit with a hammer the damping capacity can be roughly assessed by the sound of the note emitted. This method is obviously limited to gross defects and considerable skill and experience are necessary if the results are to be reliable. The method is not suitable for rigid structures due to insufficient resonance.

(ii) *Ultrasonic Testing*

This method is based upon the principle that the transmission of a beam of ultrasonic waves is upset by the presence of defects in their path. One technique is the 'echo-reflection' method which in modern flaw detectors uses a single probe combining the duties of transmitter and receiver, with the cathode ray oscillograph technique to record the indications. Fig. 15.5(*a*) shows the indications obtained when the metal is free from defects. When a defect is present some energy pulses return to the probe before those from the boundary and give rise to the indication shown in Fig. 15.5(*b*).

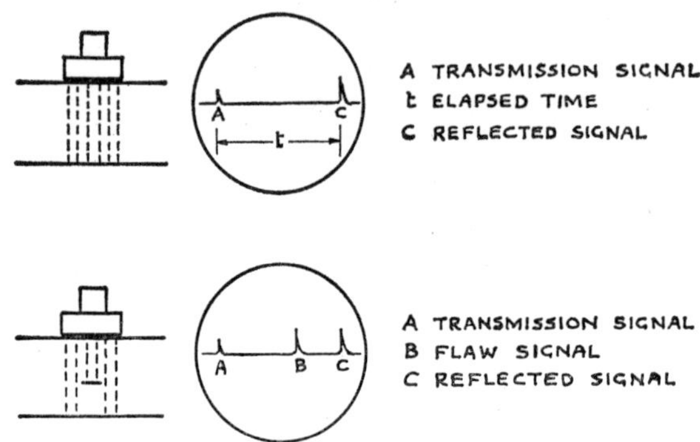

Fig. 15.5. One Method of Ultrasonic Flaw Detection
(*a*) Metal Free from Defects
(*b*) Presence of Flaw in Metal
From *Ultrasonic Flaw Detector Mark 5* (Kelvin and Hughes)

Ultrasonic testing is employed for the inspection of forgings, extrusions, castings, bar stock, welded pressure vessels and rolls.

The interpretation of the instrument indications is difficult and here again considerable skill and experience are required.

16. The Mechanical Properties of Metals and Alloys

TENSILE PROPERTIES

The Tensile Test

One of the most widely used mechanical tests is the tensile test. The test provides certain data such as the tensile strength, yield point, percentage elongation, percentage reduction of area, and the modulus of elasticity. In addition it indicates the extent of elastic and plastic deformation and so provides a measure of toughness.

The tensile properties obtained depend to some extent on the size and shape of the test specimen. Standardised specimens are therefore used. Details of the standard specimens and the method of testing are given in BS 18.1962 'Methods for Tensile Testing of Metals'.

The Load-extension Diagram for Mild Steel

When a mild steel test piece is tested in tension the relationship between load and extension will generally be of the form shown in Fig. 16.1. Over the range OA the extension is proportional to the load. The point A, where the graph deviates from a straight line, is known as the *limit of proportionability*. With mild steel this point practically corresponds to the *elastic limit,* which is the point up to which the material remains elastic. Within the elastic limit the test piece will return to its original dimensions upon the removal of the load. The coincidence of these two points is not generally true for other materials or for materials that have been overstrained. Accurate measurements show that the graph ceases to be straight before the elastic limit is reached. When the elastic limit has been exceeded the extension is permanent and is referred to as plastic deformation.

Slightly above the limit of proportionability the *yield point,* B, is obtained when a sudden permanent extension, BC, occurs without further increase in load. Sometimes there is a slight drop in the load at the yield point giving an upper and lower yield point. After the maximum load D has been passed the cross-sectional area becomes noticeably smaller and a 'neck' is formed, with the result that an increase in stress can be obtained with a reduction of load. The actual breaking load E is therefore less than D although the actual stress at E is greater than that at D.

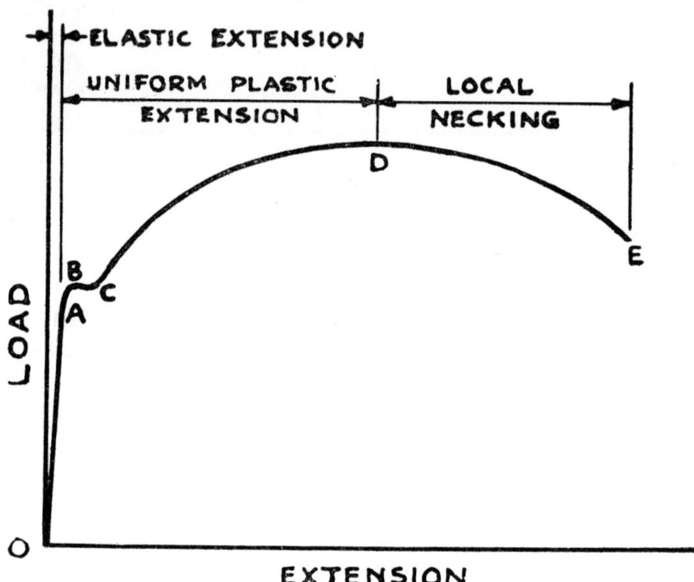

Fig. 16.1. Load-extension Diagram for Mild Steel

The results of the tensile tests may be expressed as follows:

(i) Yield Point Stress $= \dfrac{\text{Yield Load}}{\text{Original Cross-sectional Area}}$

(ii) Tensile Strength $= \dfrac{\text{Maximum Load}}{\text{Original Cross-sectional Area}}$

(iii) % Elongation $= \dfrac{\text{Increase in Length of Gauge Length}}{\text{Original Gauge Length}} \times 100$

(iv) % Reduction of Area $= \dfrac{\text{Decrease in Cross-sectional Area}}{\text{Original Cross-sectional Area}} \times 100$

Proof Stress

In metals and alloys where there is no well-defined yield point such as non-ferrous metals and alloys, and the harder steels, the term *proof stress* is used. The proof stress is the stress required to produce a permanent extension equal to a specified percentage of the gauge length, thus, using a 50 mm gauge length, the 0·1% proof stress is the stress required to produce a permanent extension of 0·1% of 50 mm, namely 0·05 mm. This can be obtained by marking off a point M (Fig. 16.2) at a distance 0·05 mm from the origin O, and

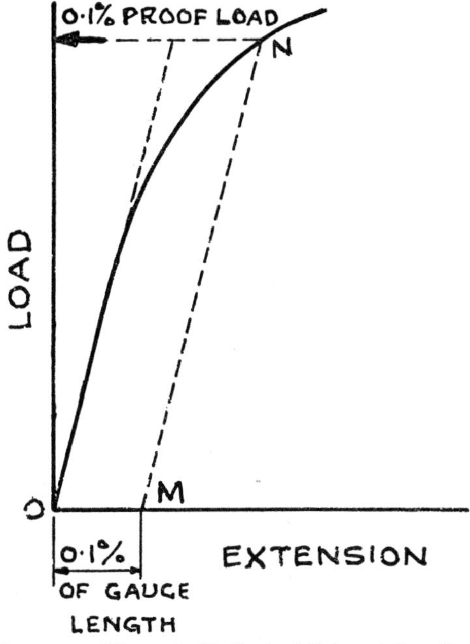

Fig. 16.2. Diagram to Illustrate Method of Determining the 0·1% Proof Stress

drawing a line MN parallel to the straight line portion of the graph which intersects the curve at N. The load corresponding to the point is the 0·1% Proof Load. If this load is divided by the original cross-sectional area the 0·1% Proof Stress is obtained.

Tensile Test Fractures

In general it is possible to distinguish between fractures that occur after the load-extension curve has passed its peak (for ductile failure see Fig. 16.3(*c*)), and fracture without a peak being reached (non-ductile failure, Fig. 16.3(*a*) and (*b*)). Type (*a*) may also be referred to as brittle failure.

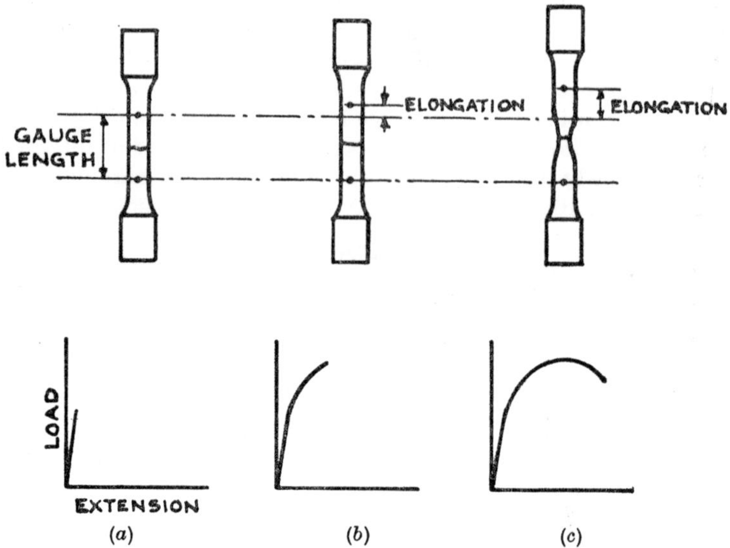

Fig. 16.3. Types of Fractures Obtained in the Tensile Test and the Corresponding Load-extension Diagrams

(*a*) Non-ductile or Brittle. No Plastic Deformation prior to Fracture No % Elongation or % Reduction of Area

(*b*) Non-ductile. Uniform Plastic Deformation prior to Fracture Low % Elongation and % Reduction of Area

(*c*) Ductile. Specimen Forms a Neck Before Fracture High % Elongation and % Reduction of Area

In (*a*) there is no plastic deformation prior to fracture, whilst in (*b*) there is uniform plastic deformation. In each case a characteristic crystalline fracture is obtained. Type (*c*) has a characteristic fibrous fracture and the load-extension diagram is typical for alloys in the soft annealed condition. However, with mild steel a well-defined yield point would be exhibited which is not shown in Fig. 16.3(*c*). With type (*c*) fracture may start at the centre of the neck and then extend outwards to give a 'cup and cone' appearance.

The area under the load-extension curve is sometimes used as a measure of the toughness of the material. This value should not be treated as a precise quantity and must not be confused with the results obtained from notched-bar impact tests. Toughness is a measure of the energy to break a material, the units therefore being force multiplied by distance e.g. Nm.

General Methods of Strengthening Metals

In Chapter 2 it was stated that plastic deformation in metals occurs by the movement of dislocations along the slip planes. Any factor which obstructs the movement of dislocations will make plastic deformation more difficult and cause an increase in the strength of the material. The various methods used for strengthening metals may be classified as follows:

(i) Grain refinement
(ii) Alloying
(iii) Work hardening
(iv) Quench hardening
(v) Precipitation or age hardening.

1. Grain Refinement

Since the slip planes in any grain have a different orientation to those in a neighbouring grain, the grain boundary will provide an obstacle to the passage of dislocations. A pile up of dislocations will occur at the grain boundary. The finer the grain size the more grain boundaries there are to produce such obstacles. Grain refining treatments will therefore result in an increase in strength.

2. Alloying

The introduction of an alloying element may produce different types of alloy structure and it is therefore difficult to generalise. However, when a solid solution is formed the solute atoms will be of different size to those of the parent metal. This results in lattice distortion which prevents the movement of dislocations.

3. Work Hardening

This is the only method of increasing the hardness of pure metals (page 12) and of many non-ferrous alloys of the single solid solution type, such as 70:30 brass and the cupro-nickels.

4. Quench Hardening

This is a characteristic of alloys with a eutectoid type of diagram, see plain carbon steels (page 47) and 10% aluminium-bronze (page 104).

5. Precipitation Hardening

Alloys which are able to be age-hardened or precipitation-hardened have a falling solid solubility curve (Fig. 11.1). The phenomenon is discussed in connection with alloys of the Duralumin type on page 110. Other alloys which are strengthened in this way include certain of the Nimonic alloys and stainless steels, Beryllium-Copper, and K-Monel.

HARDNESS

The term hardness is a difficult property to define since it can refer to resistance to indentation, abrasion, machining or scratching. That these forms of hardness are different is illustrated by Hadfields Manganese Steel (page 70) which has a low indentation hardness yet possesses a good wear and abrasion resistance. Engineers have usually accepted indentation hardness as their basis of hardness measurement. The results of such tests are useful as a control in heat treatment operations and are related to the yield strength and the tensile strength of the material.

Four hardness tests will be considered—namely, the Brinell, Vickers, Rockwell and the Shore Scleroscope.

1. Brinell Hardness Test

This test was introduced by Brinell in 1901 and is covered by BS 240.1962.

A hardened steel ball is pressed into the surface of the specimen for 10–15 seconds, and the diameter of the impression is measured using a low-power graduated microscope.

The Brinell hardness number, denoted by the symbol HB, is obtained by dividing the load in kilogrammes-force by the spherical area of the impression in square millimetres.

$$HB = \frac{2F}{\pi D[D - \sqrt{D^2 - d^2}]}$$

where F = load in kgf

D = diameter of the ball in millimetres.

d = mean diameter of the indentation in millimetres

For a given value of F and D the Brinell hardness number corresponding to the diameter of the impression may be read off from tables.

It is found that the Brinell hardness is not constant for a given specimen but depends upon the load and the size of the ball. It is therefore necessary to adopt a standard value of $\frac{F}{D^2}$ depending upon the material (Table 16.1).

When expressing the Brinell hardness number, the symbol HB is supplemented by numbers indicating the diameter of the ball and the load applied. Thus 250 HB 10/3000 indicates that a Brinell hardness of 250 was obtained using a 10 mm diameter ball and a load of 3000 kgf.

The Brinell test is unsuitable for very hard materials due to deformation of the steel ball indentor. This deformation results in indentations of greater diameter than those that would have been obtained if the ball had not been deformed. The resulting hardness numbers are therefore lower than the true hardness values. The error will increase with increasing hardness of the material being tested, which means that the Brinell hardness scale is not proportional, thus, although a material of 250 HB is twice as hard as one of 125 HB, a material of 800 HB is not four times as hard as one of 200 HB.

Diameter of Ball mm	*Load kgf*			
	Steels and Cast Irons	*Copper and its Alloys Aluminium Alloys*	*Aluminium*	*Lead, Tin, and their Alloys*
	$\frac{F}{D^2}=30$	$\frac{F}{D^2}=10$	$\frac{F}{D^2}=5$	$\frac{F}{D^2}=1$
1	30	10	5	1
2	120	40	20	4
5	750	250	125	25
10	3 000	1 000	500	100

Table 16.1. Correlated Values of Load and Ball Diameter for Brinell Hardness Test on Various Materials

For metals of hardness greater than 450 HB a tungsten carbide ball is recommended. The resulting values are not true Brinell hardness numbers and such tests are regarded as 'modified' Brinell tests.

The Brinell machine produces a relatively large indentation, which is desirable when it is necessary to obtain the average hardness of an heterogenous material like grey cast iron. However, it will be unsuitable for thin sheet metal, plated and surface hardened metals.

The Brinell hardness number can be used to estimate the approximate tensile strength (N/mm^2) of steels from the relationship:

$$TS = 3{\cdot}4.HB$$

This relationship is not true for severely cold worked or austenitic steels.

2. The Vickers Hardness Test

This method was introduced by Smith and Sandland in 1922 and developed by Vickers Armstrong Ltd. It is covered by BS 427.1961.

The indentation is made using a square base diamond pyramid with an angle of 136° between opposite faces. The loads vary from 5 to 120 kgf depending upon the material and are usually applied for 15 seconds.

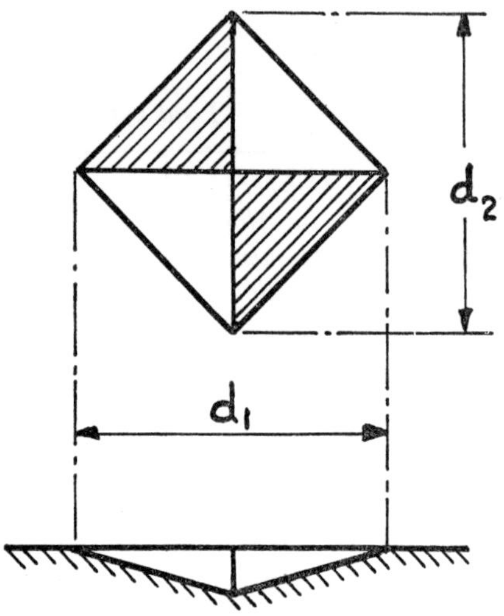

Fig. 16.4 Dimensions of Diamond Indentation

The Vickers hardness number, denoted by the symbol HV, is the load in kgf divided by the contact area of the impression in square millimetres.

$$HV = \frac{2F \sin \frac{136^\circ}{2}}{d^2} = 1{\cdot}854 \frac{F}{d^2} \text{ approx.}$$

where F = load in kilogrammes force.
d = arithmetic mean of the 2 diagonals d_1 and d_2 in millimetres.

The symbol **HV** is usually supplemented by a number indicating the load used in the test. Thus 650 HV 30 indicates a Vickers hardness value of 650 obtained using a 30 kgf load.

Since all impressions are geometrically similar the hardness values obtained are independent of the load. The hardness scale is truly proportional, so a material of 800 HV is eight times as hard as one having a hardness of 100 HV. It thus provides a rational scale of hardness from the softest to the hardest material. For low hardness values, up to about 300, the Vickers and Brinell hardness results are approximately the same, but at higher values the Brinell results are lower due to distortion of the ball indentor.

3. The Rockwell Hardness Test

This test was introduced by Rockwell in 1922 and is covered by BS 891.1962.

The Rockwell hardness number is a measure of the depth of penetration of a standard indentor. Nine scales of hardness are available (A to K inclusive) but the ones usually employed are the B and C scales, details of which are given in the following table:

Scale	*Indentor*	*Load kgf*		
		Minor	*Major*	*Total*
B	1·6 mm dia. steel ball	10	90	100
C	120° diamond cone	10	140	150

A minor load of 10 kgf is first applied and the dial indicator set to zero. The load is then increased to 100 kgf (B scale) or to 150 kgf (C scale). When the reading of the dial indicator is steady the major load is taken off and the hardness number read off directly from the dial indicator.

The principle of the method is illustrated in Fig. 16.5.

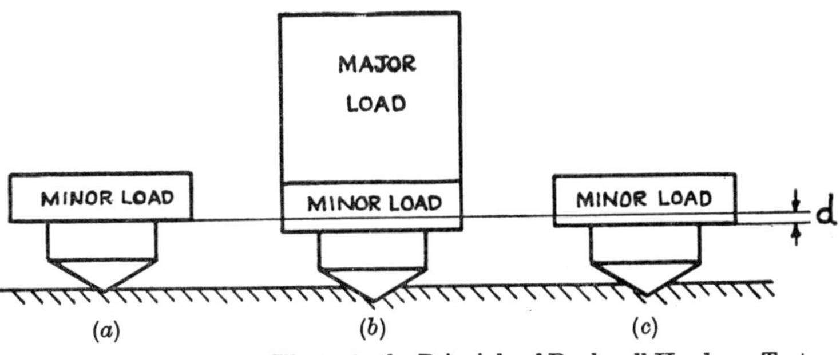

Fig. 16.5. Diagrams to Illustrate the Principle of Rockwell Hardness Test
(*a*) Minor Load Applied
(*b*) Major Load Applied
(*c*) Major Load Removed

The dial indicator records a measure of the depth increment d mm due to the major load. The dial is divided into 100 divisions each representing 1 point on the hardness scale and corresponding to a penetration of 0·002 mm. The relationship between the hardness number on each scale and the depth increment d is as follows:

$$\text{HRC} = 100 - \frac{\text{d}}{0{\cdot}002}$$

$$\text{HRB} = 130 - \frac{\text{d}}{0{\cdot}002}$$

The Rockwell hardness scale is not proportional. Reference to Fig. 16.6. shows that for low hardness values an increase of 1 HRC corresponds to approximately 5 HV whilst at high hardness values

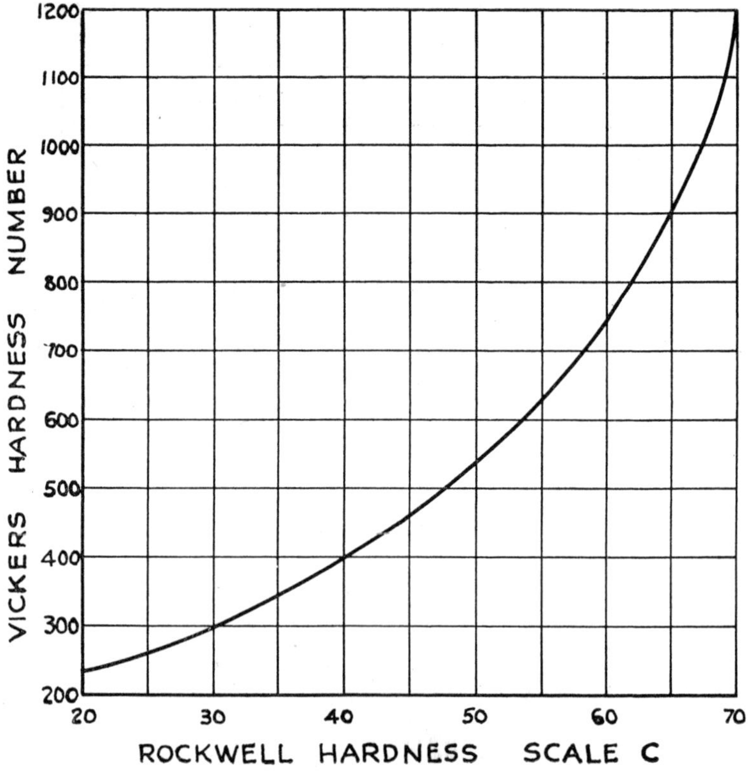

Fig. 16.6. Graph to Illustrate Relationship between Vickers Hardness Number and Rockwell Hardness Number (Scale C)

an increase of 1 HRC may correspond to approximately 40 HV. The Rockwell scale is therefore not so accurate as the Vickers scale in distinguishing between small differences in hardness of hard materials. However, the test is particularly suitable for rapid routine checks on the hardness of finished products.

4. The Shore Scleroscope

The Shore Scleroscope provides a dynamic hardness test in which a small diamond-pointed hammer of mass 2·36 g is allowed to fall freely from a height of 254 mm on to the test piece. The height of rebound is measured against a scale graduated into 140 equal divisions. A higher rebound is obtained with hard metals than with softer metals.

The advantages of this test are its portability and the fact that it leaves no visible impression. It is particularly suitable for large rolls, gears, dies and castings, and for the testing of rubber.

TOUGHNESS AND BRITTLENESS IN METALS

The terms toughness and brittleness are frequently used but are difficult to define with accuracy since there are so many anomalies and contradictions.

The amount of plastic deformation prior to fracture may be used to distinguish between toughness and brittleness. If fracture occurs with little or no prior plastic deformation the metal may be said to be in a brittle condition. Further significance is provided by the appearance of the fracture, a fibrous fracture being associated with toughness and a crystalline fracture with brittleness.

Many engineering structures unavoidably contain points of stress concentration and the stress at these points may become sufficiently high to initiate a crack. It seems therefore more important when discussing toughness and brittleness to consider the stress required to propagate a crack once it has started. On this basis two kinds of fracture may be distinguished—namely, *stable* and *unstable*, the former being associated with toughness and the latter with brittleness.

In the *stable* kind of fracture the separation is gradual and can be controlled by the external load. The crack, once formed, requires a lot of plastic deformation at its tip to maintain propagation and will stop when it travels from a region of high stress to one of normal stress.

In the *unstable* kind of fracture the separation is sudden and progresses under the influence of stored elastic energy, no external energy being necessary for propagation. The fracture is uncontrollable and the crack, once started, continues to propagate even when it moves into a region of normal stress. The crack accelerates rapidly to a velocity approaching that of sound.

Toughness and brittleness considered in this context are often described as notch-toughness and notch-brittleness, and unfortunately these are properties which cannot be actually measured. However, notched-bar tests can be used to measure the relative toughness and brittleness of the metal under the conditions of the test.

The Izod Notched-bar Impact Test

Many notched-bar impact tests are available but probably the one most widely used in this country is the Izod test.

The Izod impact value is the energy in joules required to fracture a standard notched specimen. The standard test piecemay be of square section (10 mm × 10 mm) or of circular cross-section (11·4 mm diameter) and may contain one, two or three notches. The latter is more usual so that three measurements of energy can be made from each specimen.

The dimensions of a round Izod test specimen are shown in Fig. 16.7(*a*).

The specimen is tested as a cantilever, being gripped in a vice such that the notch is in the plane of fixing. A hammer, in the form of a swinging pendulum fitted with a horizontal knife edge, is

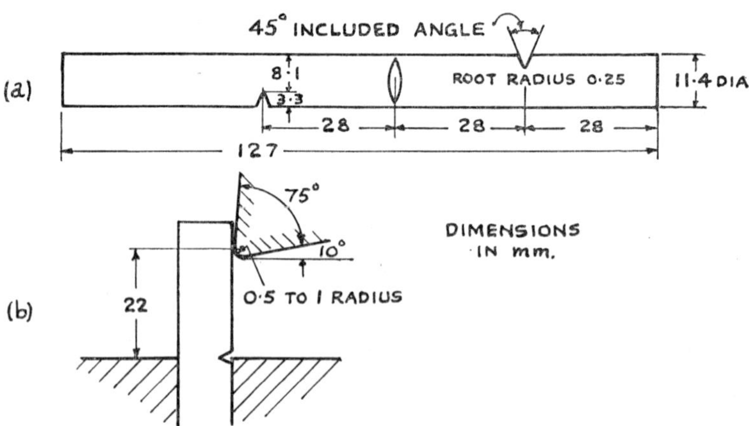

Fig. 16.7. (*a*) Dimensions of Round Izod Test Specimen
(*b*) Method of Fixing and Position of Striking of Izod Test Specimen

released from a fixed height and hits the specimen at a point 22 mm above the notch as shown in Fig. 16.7(*b*). The pendulum will continue to swing after fracturing the specimen and the height which it reaches is a measure of the energy absorbed in fracture. If the specimen is brittle very little energy will be absorbed in fracture and the pendulum will rise to a greater height than that reached when a 'tough' specimen is used. The pendulum actuates a loose pointer which moves over a calibrated scale and the energy absorbed in fracture can therefore be read off directly. The striking energy is usually 163 joules and the speed of the striker at the instant of impact is 3·5 metres per second. A value of 3 to 4 m/s is quoted in BS 131, Part 1.1961.

Significance of the Izod Test Results

It will be apparent that there are three factors present in the Izod test which are absent in the tensile test—namely, the presence of a notch, the bending, and the impact. Of these the most important is the presence of the notch which introduces multi-axial stresses.

The Izod test results have no design significance. Two materials of equal Izod value, under a certain set of conditions, may give quite different results if the specimen size, dimensions of the notch, or the temperature are varied. The appearance of the fracture probably provides as much information as the actual Izod value. The test does not measure the shock resistance of the material since the speed of testing is too low.

Despite these limitations the Izod test can be used to:

(i) Provide an indication of incorrect heat-treatment. Temper-brittleness in nickel-chrome steels (page 75) is only revealed by notched-bar impact tests. The results of the tensile test do not discriminate between the steels in this case.
(ii) Indicate the behaviour of the material to multi-axial stressing under the conditions of the test.
(iii) Study the factors which cause notch-brittle fracture in mild steel.

Notch-brittle Fracture in Mild Steel

Mild steel is usually considered to be a ductile material, but unfortunately under certain conditions it can fail in a brittle manner. When such brittle fractures have occurred in service the results have been disastrous. Well known examples of brittle fracture are the breaking in two of some all-welded Liberty ships during World War II and the collapse of an all-welded steel bridge at Hasselt in

Belgium over thirty years ago. Other serious fractures have occurred in large oil-storage tanks and welded pipe lines.

Although the use of welding in large-scale construction has made the problem more prominent, it should not be assumed that it only occurs in welded construction. With riveted construction a brittle crack once started will run to the edge of the plate. A new crack would have to be initiated before the next plate would fail. However, in welded construction the crack has a continuous path through the whole structure, giving more serious results.

Notched-bar impact tests by introducing multi-axial stresses can in some cases cause brittle fracture and have therefore been used to study this phenomenon.

Transition Temperature

The results of Izod tests on mild steel specimens carried out over a range of temperature show that as the temperature is decreased there is a fairly sudden change from high Izod values and fibrous fractures to low Izod values and crystalline fractures. This change takes place over a range of temperature known as the *transition range* and is illustrated in Fig. 16.8.

This transition range does not appear to exist for metals which

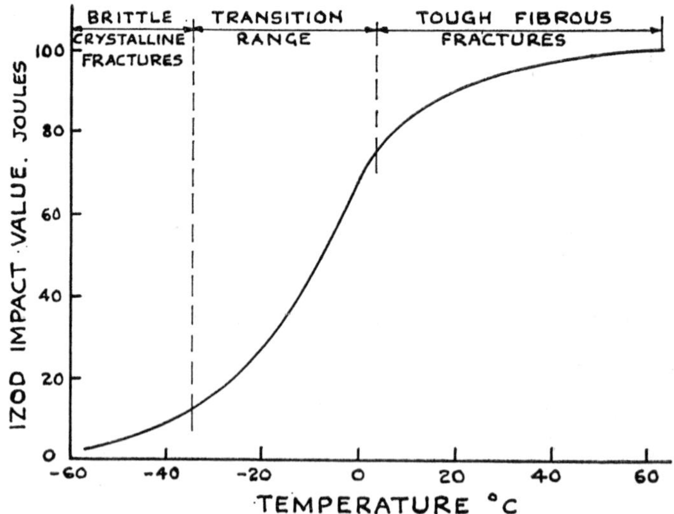

Fig. 16.8. Effect of Temperature on the Izod Impact Value of a Mild Steel Specimen

possess a face-centred cubic lattice, such as copper, aluminium and the austenitic steels. It is found mainly in ferritic steels.

A great deal of fundamental work is being carried out on the problem of brittle fracture. It seems apparent that to prevent the initiation of a crack and its spread in a brittle manner it is necessary to lower the transition temperature as far as possible.

Factors Affecting the Transition Temperature

1. Chemical Composition

Carbon, silicon, phosphorus and sulphur raise the transition temperature, whereas nickel and manganese lower it.

The resistance to brittle fracture in structural mild steels can be improved by increasing the Mn:C ratio as in the ND series of steels (BS 2762.1956).

The increased use of cryogenic equipment for holding and transporting liquefied gases, such as nitrogen, has led to the development of the 9% nickel steel. This alloy has a good tensile strength and resistance to brittle fracture at temperatures down to $-200°C$.

2. Grain Size

A decrease in grain size lowers the transition temperature. Treatments which result in grain refinement are therefore beneficial in avoiding brittleness.

3. Thickness of Hot Rolled Plate

It is found that thick plates have higher transition temperatures than thinner plates that have been hot rolled from the same ingot. There is therefore a greater risk of brittle fracture in heavy sections. This is probably because a higher finishing temperature is employed in the hot rolling of thicker plates resulting in a coarser grain size.

4. Work Hardening

Local or general plastic deformation during fabrication or in service may result in work hardening which will increase the transition temperature.

5. Stress Concentration

The presence of notches, whether designed or accidental, as in welding defects at critical and highly stressed locations, can increase the tendency to brittle failure. The sharper the notch the higher the transition temperature.

6. Impact

Impact or sudden loading increases the tendency to brittle fracture.

However, many of the service failures have occurred under essentially static conditions.

7. Neutron Irradiation

Neutron irradiation raises the transition temperature with increasing dose and reduces the energy absorbed in the temperature range in which tough fractures occur.

Most of the service failures due to brittle fracture have occurred in welded structures at moderately low stresses well within the designed limits. The fractures have occurred in or near welds, but in most cases did not follow the weld. Most fractures have had the characteristics of unstable fracture as described earlier. Low temperature is probably the most important factor, most of the fractures having occurred in cold weather.

It should however be borne in mind that although such failures have serious results the number that have occurred has been relatively small.

CREEP

The term *creep* is used to describe the slow *plastic* deformation that occurs under prolonged loading, usually at high temperatures.

The study of creep is of great concern in:

(i) the design of gas turbines,
(ii) aircraft and rockets where the material is subject to aerodynamic heating in high-speed flight, and
(iii) oil, steam and chemical plant operating at high temperatures.

At room temperatures creep may occur in lead pipes and roofing, and in white metal bearings.

Creep can be studied by submitting specimens to a constant load and temperature and measuring the permanent extension after various time intervals. If the results are plotted a creep curve of the type shown in Fig. 16.9 will be obtained.

After an initial instantaneous extension, the period of creep to the point of fracture is usually defined in terms of three stages, as follows:

(i) The *primary* stage, AB, in which, as a rule, strain hardening is taking place and there is a relatively large rate of strain which decreases with time.
(ii) The *secondary* stage, BC, in which the rate of strain is approximately constant and has its lowest value. The secondary stage is the most important part of the curve and should extend to cover the whole of the estimated service life of the alloy.

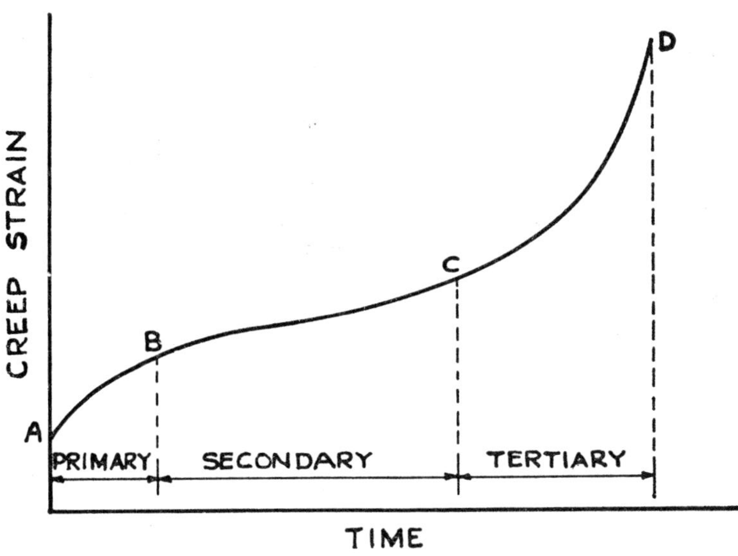

Fig. 16.9. Typical Creep Curve Showing the Three Stages of Creep

(iii) The *tertiary* stage, CD, in which the strain rate accelerates sharply and finally fracture occurs at D.

The distinction between the three stages is not very clear in many cases. The secondary stage can be regarded as the result of a precarious balance between strain hardening and the softening effect due to recrystallisation at high temperature.

A creep curve of the type shown in Fig. 16.9 represents the behaviour of an alloy at a particular temperature and load. To obtain a complete picture of the creep properties of the alloy it is necessary to construct creep curves for a range of stresses over a range of temperatures. However, creep design data are usually presented as a series of curves relating either:

(i) stress and time for various amounts of strain or rupture at a given temperature Fig. 16.10(*a*).
(ii) stress and temperature for various times to produce a given amount of strain Fig. 16.10(*b*).

Most of the creep strain appears to be due to ordinary crystal slip. Grain boundary flow also occurs at high temperature, but this phenomenon is not so important in the stages of creep of practical interest to the engineer.

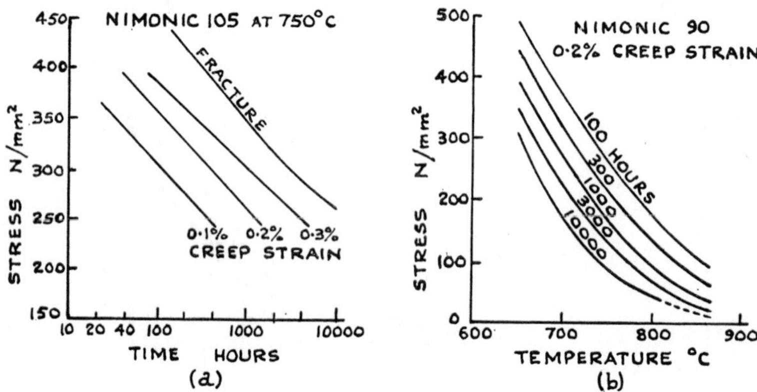

Fig. 16.10. (*a*) Stress to Produce Definite Amounts of Creep Strain in Various Times at 750°C. for Nimonic 105

(*b*) Stress to Produce 0·2% Creep Strain in Definite Times at Various Temperatures for Nimonic 90

(Redrawn by permission of Henry Wiggin and Co. Ltd. and converted to S. I. units from their publication No. 2358, The *Nimonic Series of High Temperature Alloys.*)

The Characteristics of Creep-resisting Alloys

Since creep involves strain hardening it is clear that any structural factor that opposes the movement of dislocations will reduce the creep rate. Creep-resisting alloys are therefore of the type which contain obstacles, such as finely dispersed particles, which block the passage of dislocations. These are usually produced by precipitation-hardening treatments and the particles should not coagulate or dissolve at the service temperature. Since a certain amount of grain boundary flow is involved, coarse-grained structures have a better creep resistance than fine-grain structures.

For use in steam-power plant low-alloy steels (page 82) containing 0·5% molybdenum and other carbide-forming elements are suitable. For use at higher temperatures, such as gas turbines, other properties such as resistance to oxidation and scaling, become important (page 81), for which the Nimonic alloys are particularly suitable (page 119). Many precipitation-hardening stainless steels, like 17/7 PH (17Cr–7Ni–Al) and 17/4 PH (17Cr–4Ni–Cu–Nb) are also available for high-temperature applications. Many creep-resisting alloys are in the development stage.

FATIGUE OF METALS

The term *fatigue* was introduced by Braithwaite in 1854 to explain the failure of metals under repeated applications of a load which it could sustain indefinitely when applied statically.

The crystalline nature of the fatigue fractures led early workers to believe that the metal developed a crystalline structure under repeated stressing. However, this idea was disproved with the introduction of microscopic examination of metals, when it was shown that all metals were composed of crystalline grains. The characteristic appearance of fatigue fractures is discussed on page 150.

Since it is believed that about 90% of the service failures in automobiles and aircraft components are due to fatigue, the study of this phenomenon is of great importance to the engineer.

Stress Cycles

The repeated variation in stress involved in fatigue is referred to as a stress cycle. The terms used in connection with stress cycles are illustrated in Table 16.2, where tensile stresses are taken as positive and compressive stresses as negative.

Example	*Maximum Stress* S_{max}	*Minimum Stress* S_{min}	*Range of Stress* $2\,S_a$	*Stress Amplitude* S_a	*Mean Stress* S_m
(i)	+120	−120	240	120	0
(ii)	+ 90	− 60	150	75	+15
(iii)	+150	+ 30	120	60	+90

Table 16.2. Examples of Stress Cycles

Example (i) consists of a completely reverse stress of equal magnitude whilst (ii) represents one of unequal magnitude. In example (iii) the stress is entirely tensile but of variable magnitude.

Many types of repeated loading may give rise to fatigue. The examples given below are taken from BS 3518, Part I. 1962, *Methods of Fatigue Testing* to which reference should be made:

(i) Direct stress, e.g. the loading on a piston rod.
(ii) Plane bending, e.g. the loading applied to leaf springs.
(iii) Rotating bending, e.g. the loading on the rotating axle of a railway wagon.
(iv) Torsion, e.g. a vehicle suspension torsion bar.
(v) Combined stresses, e.g. the loading on a crankshaft journal, which is a combination of bending and torsion.

Fatigue Testing

There are many types of fatigue tests and the reader is referred to BS 3518, *Methods of Fatigue Testing*, Parts I and II. 1962 and Part III. 1963.

The most commonly used type of laboratory fatigue test is that of the rotating bending type (Fig. 16.11). In this type of test the

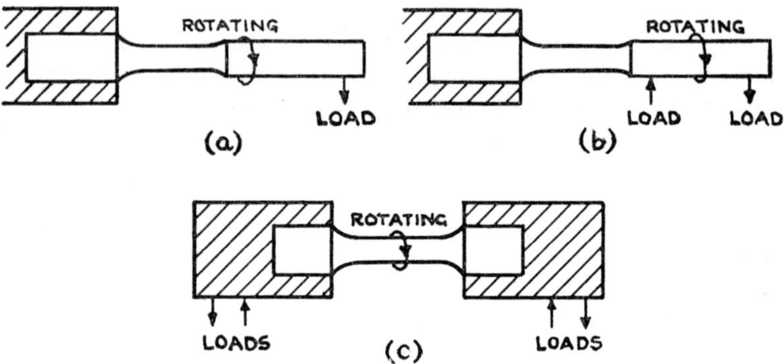

Fig. 16.11. The Rotating Bending Fatigue Test Using a Cylindrical Specimen

(*a*) Single-point Loading
(*b*) Two-point Loading
(*c*) Four-point Loading

outer fibres of the test piece of circular cross-section suffer alternate tension and compression as the specimen rotates.

With single-point loading Fig. 16.11(*a*) there is a direct shear on the section in addition to the bending stress. This can be overcome by the use of two-point or four-point loading, as in Fig. 16.11(*b*) and (*c*). These methods give a constant bending moment and zero shear force over the gauge length of the specimen. The gauge length may also be tapered or toroidal in order to give a more uniform distribution of stress.

The fatigue resistance of a metal can be determined by taking a number of specimens and testing them at various stresses. The number of reversals of stress required to fracture the specimen with each range of stress is noted. The range of stress is progressively reduced with each specimen until eventually a stress range is reached at which fracture does not occur after 10^7 to 10^8 reversals. A graph is then plotted of the stress amplitude S against the logarithm of the number N of stress cycles, to give what is usually known as the S–N curve for that metal, shown in Fig. 16.12.

It will be noticed from Fig. 16.12(*a*) that below a certain stress, known as the *fatigue limit*, fracture will not occur even if the stress is repeated an infinite number of times. The fatigue limit in steels is reached after about 10^6 reversals. However, with non-ferrous alloys and austenitic steels no definite fatigue limit is obtained as in Fig. 16.12(*b*). In such cases it is usual to quote an *endurance limit*, which is the stress required to produce fracture after a definite number of stress cycles.

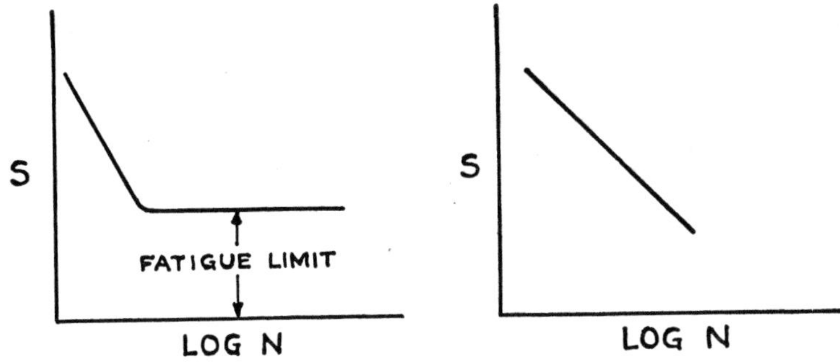

Fig. 16.12. Typical S–N Curves for
(*a*) a Plain Carbon Steel
(*b*) a Non-ferrous Alloy

Although laboratory tests on small metal specimens provide much useful information, a more reliable forecast of the behaviour of a component or structural member in service is obtained by testing it under simulated conditions.

Factors Affecting the Fatigue Limit

1. The Mean Stress

Unlike many engineering applications the mean stress in fatigue tests of the rotating-bending type is zero. Empirical rules have therefore been suggested to enable the engineer to estimate the safe stress amplitude for some other mean stress value. Three such relationships are those due to Gerber, Goodman and Soderberg which may be represented as follows:

GERBER'S LAW $$S_a = S_F\left[1 - \left(\frac{S_m}{S_u}\right)^2\right]$$

MODIFIED GOODMAN LAW $$S_a = S_F\left(1 - \frac{S_m}{S_u}\right)$$

SODERBERG'S LAW $$S_a = S_F\left(1 - \frac{S_m}{S_y}\right)$$

where S_a = the stress amplitude associated with a mean stress S_m
S_F = alternating fatigue strength
S_u = tensile strength of the metal
S_y = yield strength of the metal

Such relationships can be represented graphically as shown in Fig. 16.13.

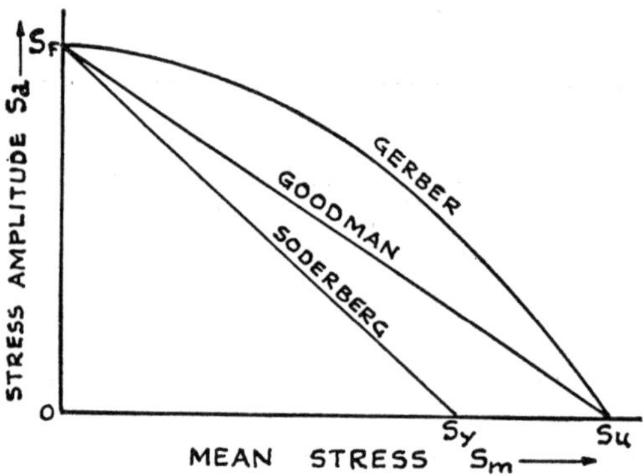

Fig. 16.13. Graphs to Illustrate the Effect of Mean Stress on the Safe Stress Amplitude According to Gerber, Goodman and Soderberg

According to the Gerber and Goodman relationships increase in the mean stress decreases the stress amplitude which can be safely superimposed until it becomes zero at the tensile strength of the material. However, the Soderberg relationship prefers to use the yield point as the criterion of failure so that, according to this law, the safe stress amplitude becomes zero when the mean stress equals the yield point stress.

2. Design and Surface Finish

Great care should be taken in design to avoid sharp fillets, since these generally reduce the fatigue limit. Surface finish is also important, a smooth surface being essential for maximum fatigue resistance.

Certain metals and alloys are more 'notch-sensitive' than others. With ordinary grey cast iron the graphite flakes act as notches or points of stress concentration, even when the iron is in the polished condition. Soft and ductile materials are not sensitive to notches since the increased stress at the root of the notch can be readily accommodated. The less ductile materials, such as the high-tensile steels, are very 'notch-sensitive' since they cannot accommodate the increased stress. In general the fatigue limit and the 'notch-sensitivity' increase with increased tensile strength. It follows that surface finish and design are therefore particularly important in high-tensile steels.

3. Surface Treatments

Fatigue fractures almost invariably start from the surface and consequently surface treatments are important. Such treatments should increase the hardness of the surface and preferably introduce beneficial residual compressive stresses.

Surface hardening treatments include shot peening, surface rolling, case-hardening, nitriding, flame and induction hardening. Beneficial residual compressive stresses are introduced by mechanical overstraining such as surface rolling the fillets of crankshafts or scragging helical springs. Compressive stresses are also introduced by the surface hardening methods mentioned previously.

Metal coatings, although primarily employed to increase the corrosion resistance and sometimes the wear resistance, may also affect the fatigue limit. Nickel and chromium coatings introduce residual tensile stresses and are therefore detrimental to fatigue strength. Zinc coatings introduce residual compressive stresses and are therefore beneficial.

Treatments producing a soft skin such as decarburising, or cladding with a softer metal, such as Alclad, will reduce the fatigue limit.

4. The Presence of Inclusions

Inclusions at or near the surface may act as points of stress concentration and reduce the fatigue limit. This is particularly true of hard inclusions such as alumina (page 31). Soft inclusions such as manganese sulphide or lead have very little effect.

5. Temperature

The results of tests on a number of alloys show that in general the fatigue limit is increased as the temperature is decreased below that of the atmosphere. It is difficult to generalise on the effect of increasing the temperature since in some alloys structural changes can occur. In many cases a well-defined fatigue limit is not obtained in high-temperature tests.

6. Corrosive Environment

The formation of small corrosion pits (page 141) gives rise to points of stress concentration which can initiate a normal fatigue failure.

The characteristics of corrosion-fatigue are illustrated in Fig. 16.14, which shows three different S–N curves.

Curve 1 is that obtained for a metal with a normal fatigue limit. Curve 2 is for the same metal which has been previously subjected to a corrosive atmosphere prior to normal air testing. A well-defined fatigue limit is still obtained but is lower than before. Curve 3 is for the same metal tested in the corrosive atmosphere. It will be noted that there is now no fatigue limit and the resistance of the metal to fatigue is greatly reduced.

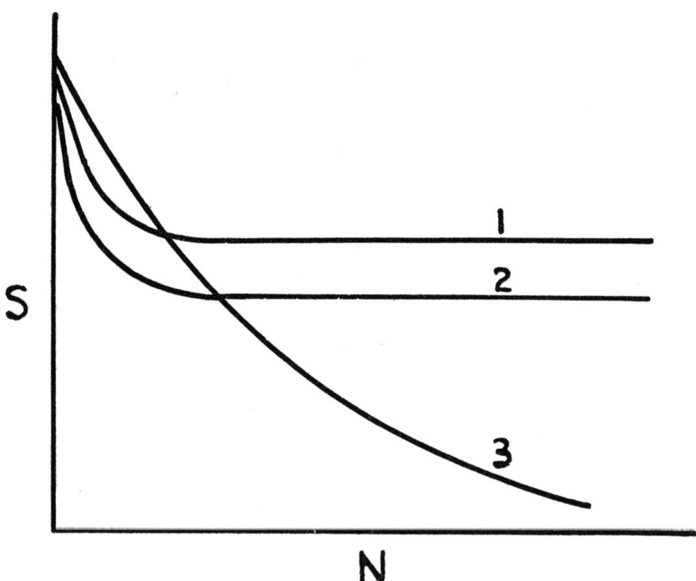

Fig. 16.14. S–N Curves to Illustrate the Effect of Corrosion Fatigue
(1) Tested in Air
(2) Tested in Air After Previously Subjected to Corrosion
(3) Tested in Corrosive Environment

Corrosion-fatigue can be prevented by care in design, the use of a more corrosion-resistant alloy or by the use of zinc coatings on steel. Zinc coatings introduce beneficial residual compressive stresses in addition to improving the corrosion resistance.

Cumulative Damage in Fatigue

In service a component may be subjected to a number of different stress amplitudes. The expected fatigue life under such conditions may be predicted from an S–N curve by means of a linear damage law known as Miner's Law. If the material is subjected to n_1 cycles of stress at a stress amplitude S_1 when the expected life is N_1 cycles then $\frac{n_1}{N_1}$ represents the fraction of life used.

If the stress is now changed to S_2 for n_2 cycles when the expected life is N_2 cycles then the fraction of life used is $\frac{n_2}{N_2}$. Miner's Law states that for any number of stress levels failure will occur when

$$\frac{n_1}{N_1} + \frac{n_2}{N_2} + \frac{n_3}{N_3} + \ldots\ldots = 1$$

In practice the value of this summation is often taken as 0·6 for safety reasons.

17. Polymer Materials

THE STRUCTURE AND PROPERTIES OF POLYMERS

Polymers include all the plastics and rubbers and they consist of long chain molecules built up by the repetition of a small simple chemical unit called a *monomer*, e.g. ethylene C_2H_4 has the molecular structure:

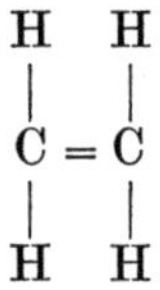

$$\begin{array}{ccc} H & & H \\ | & & | \\ C & = & C \\ | & & | \\ H & & H \end{array}$$

This monomer forms a long chain by repeatedly transferring one of its carbon bonds to give the polymer:

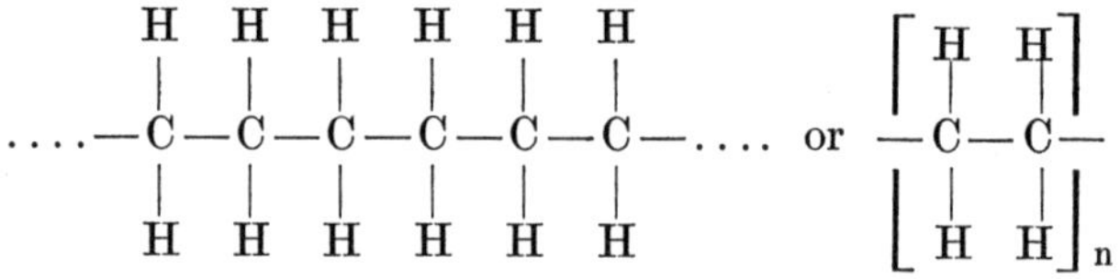

$$\begin{array}{c} \quad H \quad H \quad H \quad H \quad H \quad H \\ \quad | \quad\;\; | \quad\;\; | \quad\;\; | \quad\;\; | \quad\;\; | \\ \ldots\ldots-C-C-C-C-C-C-\ldots\ldots \\ \quad | \quad\;\; | \quad\;\; | \quad\;\; | \quad\;\; | \quad\;\; | \\ \quad H \quad H \quad H \quad H \quad H \quad H \end{array} \quad \text{or} \quad \left[\begin{array}{c} H \quad H \\ | \quad\; | \\ -C-C- \\ | \quad\; | \\ H \quad H \end{array} \right]_n$$

where — represents a pair of shared electrons known as a *covalent bond*. This process is called polymerisation and in this case yields the polymer known as polyethylene. The length of these chains may be varied but in commercial polymers chains containing from 10^3 to 10^5 of these CH_2 units are generally encountered. Since these materials are of high molecular weight, they are referred to as high polymers.

The general structural formulae of polymers such as polyvinylchloride (P.V.C.) and polytetrafluoroethylene (P.T.F.E.) may be represented as follows:

$$\left[\begin{array}{cc} H & H \\ | & | \\ -C- & C- \\ | & | \\ H & Cl \end{array}\right]_n \qquad \left[\begin{array}{cc} H & H \\ | & | \\ -C- & C- \\ | & | \\ H & F \end{array}\right]_n$$

Polyvinylchloride — Polytetrafluoroethylene

Copolymers

It is frequently possible to polymerise two monomers together so that both occur together in the same polymer chain. This normally occurs in a somewhat random fashion and the product is known as a *copolymer*. The properties of the copolymer differ from those of either component member. If A and B represent two different monomeric units, the copolymer can be represented as follows:

—A —A —B —A —A —B —B —B —A —A —B —A —

It is possible to copolymerise more than two monomers together, e.g. ABS copolymer, which consists of acrylonitrile, butadiene and styrene.

Classification of Polymers

For practical purposes polymers may be divided into thermoplastics and thermosets.

Thermoplastics soften upon heating and harden upon cooling and the whole process is capable of repetition. The structure consists of individual long chain molecules (Fig. 17.1(*a*)) and deformation can occur readily by slip between molecules against the weak Van der Waal's forces. In the solid state thermoplastics may exist as glassy amorphous solids such as polymethylmethacrylate (P.M.M.A. or 'Perspex'), or as partially crystalline materials such as nylon.

Thermosets acquire plasticity upon heating but eventually cross-linking by covalent bonds occurs between the long chain molecules (Fig. 17.1(*b*)) and the material becomes permanently set. When once cool it can no longer be softened by further heating. Thermoset scrap is no longer a moulding material. The best known example is phenol formaldehyde ('Bakelite'). Rubber is a thermoset since cross-linking occurs during curing or vulcanisation. However, it behaves mechanically like a thermoplastic.

The structure of thermoplastics and thermosets may be represented diagrammatically as in Fig. 17.1(*a*) and (*b*).

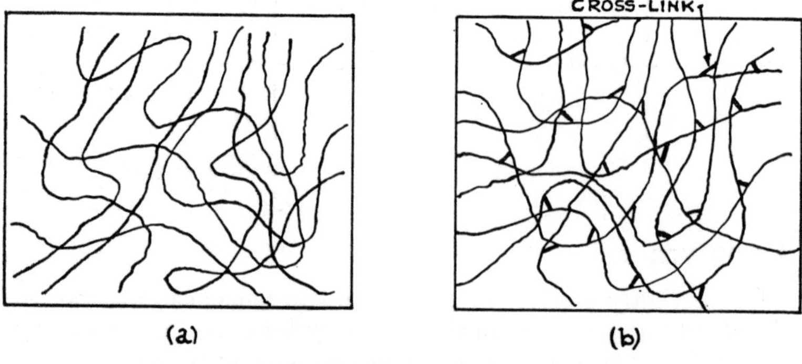

Fig. 17.1. Diagrammatic Representation of the Structure of (*a*) Thermoplastics (*b*) Thermosets

Well-known examples of thermoplastics and thermosets are given in the following table:

Thermoplastics		*Thermosets*
Amorphous	*Crystalline*	
Polystyrene ABS copolymer Acrylics P.V.C. Polycarbonate	Polyethylene Polypropylene Polyamides (Nylons) Polyacetals P.T.F.E.	Bakelite Melamine Formica Ebonite Epoxy resins

Ingredients of Commercial Plastic Materials

In addition to the resins which form the basis of the polymer, other ingredients may be present for particular purposes, e.g. plasticisers, fillers, pigments, dyes and other miscellaneous additives.

1. Plasticisers

Plasticisers are substances which, when added to a plastics material, improve the flow properties by lowering the processing temperature and increasing the flexibility. They lower the tensile strength and hardness, and for each polymer the improvement in forming properties has to be balanced against this decrease in strength.

The function of the plasticiser is to weaken the attractive Van der Waal's forces between the long chain molecules. In order to do this the plasticiser must penetrate between the molecules (i) neutralising these forces and (ii) increasing the distance between the molecules. The first function is achieved by 'primary' plasticisers, e.g. tritolyl phosphate and dioctyl pthalate for P.V.C. These do not exude from the plasticised composition after long standing or frequent flexing. 'Secondary' plasticisers have the effect of 'pushing apart' the polymer molecules, so reducing the strength of the intermolecular forces, e.g. camphor in cellulose nitrate. If used alone 'secondary' plasticisers will exude from plasticised compositions and for this reason they must be used in combination with 'primary' plasticisers.

2. Fillers

These are usually added to thermosets for economic reasons and to impart particular properties. The following fillers may be added to phenolic resins: asbestos, carbon, cotton fibre or cloth, mica, paper pulp and wood flour. Asbestos improves dimensional stability and resistance to high temperature. Carbon improves electrical conductivity whilst cotton fibre or cloth and paper pulp improve shock resistance. Mica improves dimensional stability and reduces moisture absorption. Fillers of the wood flour type prevent crazing during the curing cycle.

Metallic fillers can be used to modify the thermal and electrical properties of the polymer. Iron powder fillers reduce over-heating in tools based on resins of the epoxy type. Iron and steel powders are added to provide a basic plastic with magnetic properties. Such additions give flexible materials which provide magnetic shielding for certain electronic components, e.g. certain radar scanning components. The addition of magnetic fillers allows the forming of a magnetic material into complex shapes not readily obtainable by the use of a metal alone. A further example of the use of metallic fillers is that of powdered lead to provide flexible radiation shields in the X-ray field.

3. Pigments and Dyes

A very wide range of colours may be given to plastics by the addition of pigments and dyes. The advantage is that colour can be moulded in depth. Damage by surface abrasion therefore does not destroy the colour of the moulding.

4. Miscellaneous Additives

Numerous other additions may be made for particular purposes. Anti-oxidants retard oxidation which is a degradation mechanism in many polymers. Carbon black is known to improve the weathering properties. Antistatic additives are used to reduce dust attraction and also in films to improve the handling behaviour of certain types of bag making and packaging equipment. The addition of rubbers reduces brittleness and glass fibres improve stiffness. Other additions may be made to act as flame retarders.

THE VULCANISATION OF RUBBER

Natural rubber is a polymer made up of repeated units of the monomer isoprene. The resulting polymer, known as polyisoprene, may be represented as follows:

```
 ┌                        ┐
 │          H             │
 │          |             │
 │  H   H — C — H   H   H │
 │  |       |       |   | │
—  C   —    C   =   C — C —
 │  |                   | │
 └  H                   H ┘n
```

The long chain molecule is coiled and the process of polymer straightening under stress accounts for the high degree of elasticity. The coiled type of molecular chain is typical of a group of materials known as elastomers. Elastomers usually exhibit a reversible elongation of at least 100%.

Natural rubber can be cross-linked by heating with sulphur. This process is known as curing or 'vulcanisation'. The heat breaks some of the double bonds and the sulphur atoms link adjacent chains at these sites. If excessive cross-linking occurs the elastic properties are destroyed and a rigid structure is formed. 1% by weight of sulphur produces a soft rubber whilst 4% of sulphur produces the hard inflexible material known as Ebonite.

Cross-linking may also occur in service due to the presence of oxygen. The necessary activation energy is provided by strong sunlight or by heating. This effect can be minimised by the presence of finely divided carbon.

THE MECHANICAL PROPERTIES OF PLASTICS

When thermoplastics are subjected to a tensile test at a constant rate of strain, the stress–strain curve is not a straight line. The shape of the curve is affected by strain rate, temperature and, in many cases, the humidity. Increasing the temperature and decreasing the strain rate will reduce the strength of the thermoplastic. Humidity affects many plastics, particularly nylon, and the effect is similar to the addition of plasticiser to the material. Nearly all plastics creep at room temperatures but partially recover after the stress has been removed.

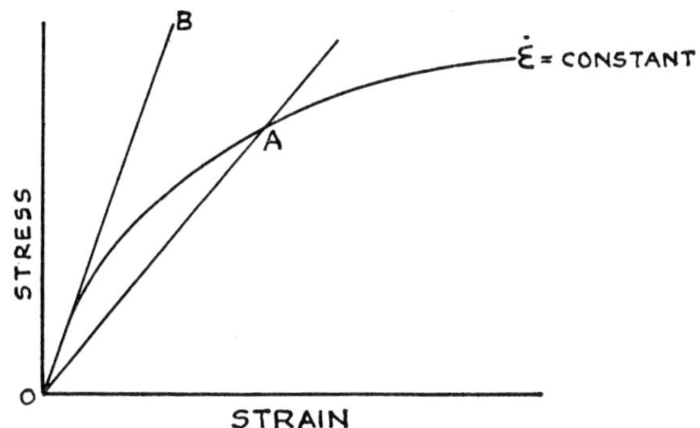

Fig. 17.2. Stress–Strain Curve for a Thermoplastic Showing Methods of Determining the Modulus

A typical stress–strain curve for constant rate of strain is shown in Fig. 17.2. Because the stress–strain curve is not a straight line the modulus can be calculated in more than one way. The slope of the line OA is called the secant modulus at the strain corresponding to the point A. The line OB is tangential to the curve at the origin O and the slope is called the tangent modulus at the origin. The tangent modulus could also be drawn corresponding to the strain at some other point on the curve. It is evident that these various definitions of modulus can give quite different results. The value in each case is also dependent upon strain rate and temperature and it is unwise to apply these data to quite different situations. Such tests are not useful for rational design since in practice the four variables stress, strain, time and temperature should be considered. Creep experiments are usually regarded as being more representative of practical situations.

FACTORS AFFECTING THE STRENGTH OF POLYMERS

1. Temperature

The long chain molecules can be considered as being in a state of constant thermal motion by rotation about each of the carbon-to-carbon bonds. At low temperatures this rotation is limited and the material is brittle and glass-like in nature. As the temperature is increased the rotations increase and the material will become more flexible.

In amorphous polymers of low molecular weight the polymer will be glassy below some given temperature whilst above that temperature it will be a viscous liquid. The melting point of such polymers will be very sharp.

With high molecular weight polymers such a clearly defined melting point does not occur. In such materials a transition occurs from glassy solid to rubbery solid. The temperature at which this transition occurs is known as the *glass transition temperature* T_g. At this transition the modulus decreases by a factor of the order of 10^3. Further heating results in a very indefinite rubber–liquid transition, sometimes referred to as the flow temperature.

The glass transition temperatures for some polymers are given below.

Polymer	*Glass transition temperature °C*
Perspex	105
Polycarbonate of bisphenol A	150
Rigid P.V.C.	74
Polystyrene	100
Natural rubber	−70
Butyl rubber	−70
Silicone rubber	−125
Neoprene rubber	−50

Most of the industrially important rubbers have very low glass transition temperatures, usually below −50°C. However, for the glass-like polymers the transition temperature is usually above 80°C.

The significance of the glass transition temperature is vividly illustrated by the behaviour of natural rubber where $T_g = -70$°C. At room temperatures it is an elastic solid but if rubber is immersed in liquid nitrogen it behaves as a very brittle solid and can be broken readily.

2. Time

At low rates of strain molecular rotations can occur readily and the material is flexible. However, at high strain rates the molecular rotations are restricted and this reduces the flexibility of the material. It can therefore be said that increasing the strain rate has the same effect as reducing the temperature.

3. Degree of Crystallisation

Some polymers are partly crystalline by their nature, i.e. they have a degree of three-dimensional order in their structure. Crystallisation may arise due to processing conditions and to elongation of the molecules of the polymer by stretching. For example, during the tensile testing of nylon orientation of the long chain molecules occurs in the necked region of the test piece. This leads to increased strength in the direction of the orientated molecules in the necked region. This increase in strength due to orientation by stretching is the basis of the strength of synthetic fibres.

In general the density, tensile strength, modulus and hardness all rise with increasing crystallinity but the impact strength falls.

4. Humidity

The tensile strength of some polymers, particularly nylon, is greatly reduced by increase in the relative humidity. For example, the tensile strength of dry Nylon 6 is reduced to 55% of its value after equilibriating the nylon at 50% relative humidity. Nylon is hygroscopic, i.e. it will absorb or lose moisture in air depending upon the relative humidity.

VISCO-ELASTICITY

The behaviour of polymers over small strains approximates to the laws of visco-elasticity. The prefix 'visco' implies that the material has some of the features of a viscous liquid so that its properties are time dependent, i.e. the response to loading depends upon the duration of application of the load.

Polymers do not exhibit the elastic behaviour of the ideal Hookean solid or the Newtonian behaviour of the ideal liquid.

The ideal solid obeys Hookes Law $\tau/\gamma = G$

where τ = shear stress
γ = shear strain
G = modulus of rigidity.

For the ideal Newtonian liquid the following relationship applies:

$$\frac{d\gamma}{dt} = \frac{\tau}{\mu}$$

where μ = coefficient of viscosity
t = time.

Flow in polymers can be represented by a combination of basic model elements such as the spring and the dashpot. The spring represents the perfect elastic solid and the dashpot represents the perfect viscous liquid, i.e. the time dependence of the behaviour. These models exhibit the same relationship between force, elongation and time that stress, strain and time do in a real material.

The simplest analogies are the Maxwell model (Fig. 17.3(*a*)) and the Kelvin model (Fig. 17.3(*b*)).

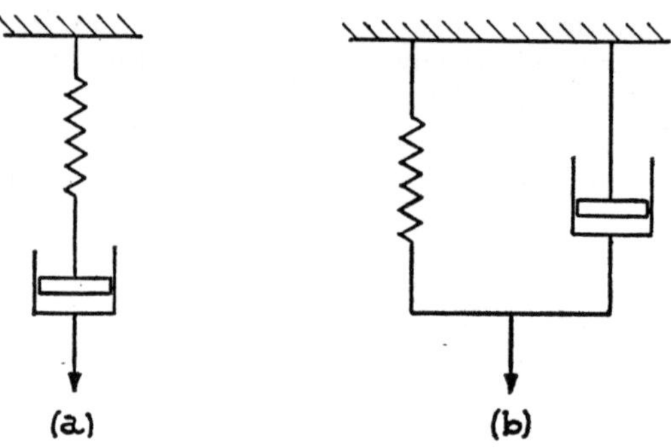

Fig. 17.3. Diagrams to Illustrate (*a*) the Maxwell Model (*b*) the Kelvin Model

In the Maxwell model the spring and dashpot are in series.

Suppose that a stress τ is applied when $t=0$ and held constant until $t=t_1$. There will be an instantaneous shear strain of $\frac{\tau}{G}$ and at any time t:

$$\gamma=\frac{\tau}{G}+\frac{\tau}{\mu}t$$

At the final time t_1:

$$\gamma_1=\frac{\tau}{G}+\frac{\tau}{\mu}t_1$$

If the stress is now removed, the elastic deformation will also be removed so that beyond $t=t_1$:

$$\gamma=\frac{\tau}{\mu}t_1$$

The results may be illustrated graphically as in Fig. 17.4.

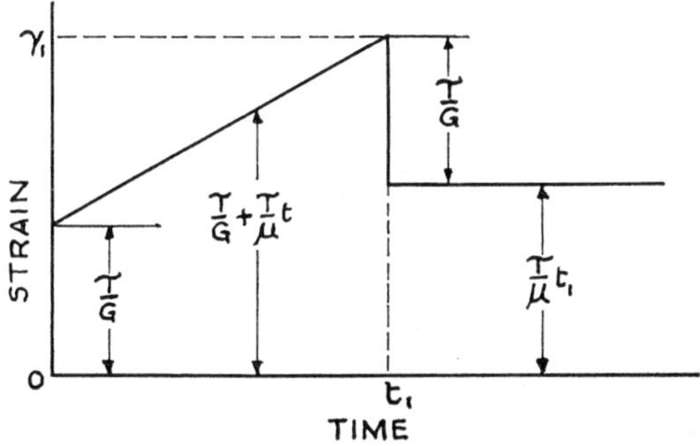

Fig. 17.4. Strain–Time Graph for Maxwell Model

When the strain is held constant the Maxwell model demonstrates how the stress relaxes with time (Fig. 17.5).

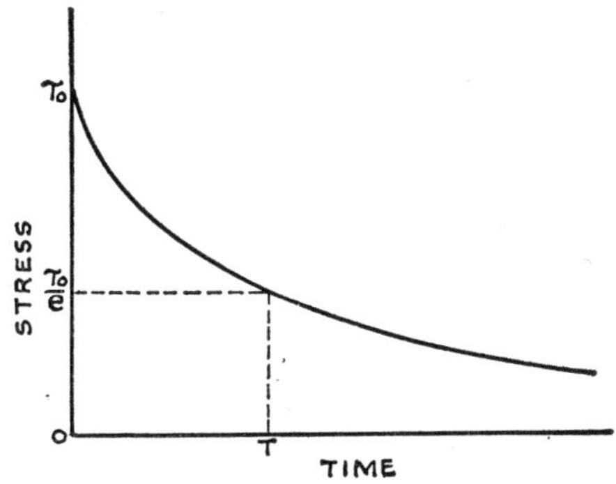

Fig. 17.5. Stress–Relaxation Graph for Maxwell Model

The time T for the stress to relax to $\frac{1}{e}$ of its original value τ_0 is known as the relaxation time. The relaxation time T can be shown to be equal to $\frac{\mu}{G}$. In the case of a polymer molecule the relaxation time provides a measure of the time taken for molecules to recoil after deformation.

In the Kelvin model the spring and the dashpot are in parallel (Fig. 17.3(*b*)).

The total stress will be the stress in the elastic element plus the stress in the viscous element.

$$\tau = G\gamma + \mu\frac{d\gamma}{dt}$$

This solves to give

$$\gamma = \frac{\tau}{G}[1 - e^{-Gt/\mu}]$$

but $\frac{\mu}{G} = T$, where T is known as the retardation time.

$$\therefore \quad \gamma = \frac{\tau}{G}[1 - e^{-t/T}]$$

If the stress is removed after time t_1, then

$$G\gamma + \mu\frac{d\gamma}{dt} = 0$$

which solves to give

$$\gamma = \gamma_{t_1} e^{-(t-t_1)/T}$$

The strain–time curve for an applied stress τ removed after time t_1 is shown in Fig. 17.6.

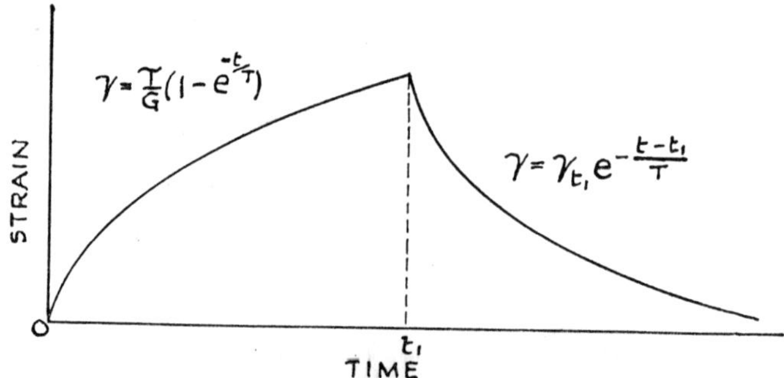

Fig. 17.6. Strain–Time Graph for Kelvin Model

The behaviour of plastics can be more accurately described by combining the Maxwell and Kelvin models in series (Fig. 17.7). This makes it possible to approximate all the deformation modes found in high polymers, i.e. the combined model allows simultaneously for instantaneous and retarded elastic deformation and for irrecoverable creep (Fig. 17.8). In this graph the applied stress is removed after time t_1.

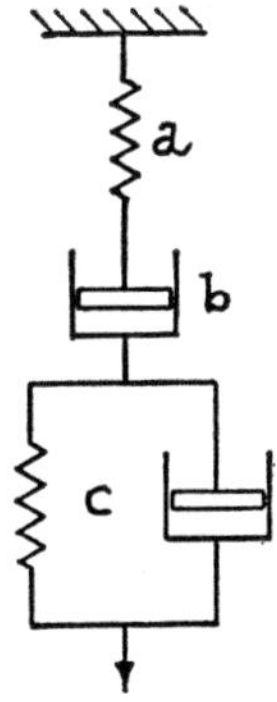

Fig. 17.7. Maxwell and Kelvin Model in Series

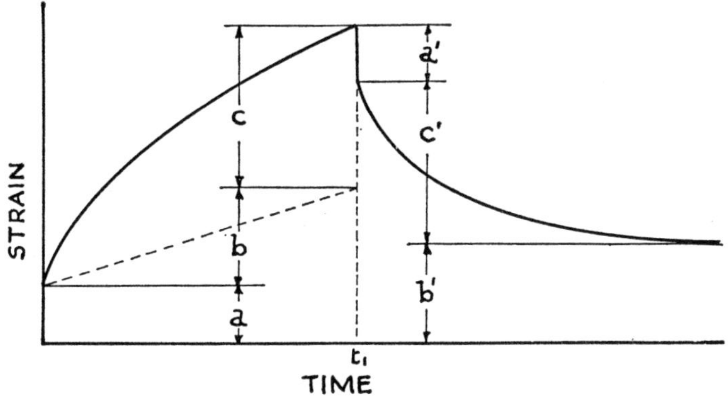

a = instantaneous spring strain
b = strain in dashpot
c = strain of spring–dashpot combination
a′ = instantaneous strain of spring
c′ = strain of spring–dashpot combination
b′ = permanent strain

Fig 17.8. Strain–Time Curve for Maxwell and Kelvin Model in series

FIBRE-REINFORCED PLASTICS

The development of fibre reinforcement was due largely to the recognition of the high strengths of polycrystalline fibres and of single crystals in the form of whiskers. This encouraged engineers to make use of these properties by binding fibres and whiskers together in a matrix to form a composite material.

The term *fibre* indicates a material of uniform cross-section which can be produced in various lengths without limit. Fibres may be made from both polycrystalline or amorphous materials and their strengths may be as high as 3 000 N/mm^2. Well-known fibres include E glass, S glass, boron, boron carbide and carbon.

The term *whisker* indicates a single crystal with growth in a definite direction. Whisker diameters vary between 0·5 to 3·0 μm and they may be up to 25 mm long. Since they are single crystals of very small cross-section, they are relatively flawless and can therefore possess almost theoretical strength. The strength of such whiskers may be as much as 30 000 N/mm^2. Examples of whiskers are alumina Al_2O_3, silicon nitride Si_3N_4, and silicon carbide SiC.

The Principles of Fibre Strengthening

In a composite material the fibres carry most of the load on the system. The plastics matrix deforms and transfers stress by means of shear tractions at the fibre-matrix interface to the fibre. If the fibres are sufficiently long they take up the same deformation as the matrix over the greater part of their length.

When the stress is applied parallel to the fibre direction, the strength and the modulus of composites containing uni-directional continuous fibre can be predicted by the law of combined action, viz.:

$$\sigma_c = \sigma_f V_f + \sigma_m V_m$$

and

$$E_c = E_f V_f + E_m V_m$$

where σ_c and E_c represent the tensile strength and modulus of the composite

σ_f and E_f represent the tensile strength and modulus of the fibre

σ_m and E_m represent the tensile strength and modulus of the matrix

V_f = volume fraction of the fibre

V_m = volume fraction of the matrix

When the applied stress is not parallel to the fibre direction it is usual to introduce an orientation factor η to the equations, viz.:

$$\sigma_c = \sigma_f V_f \eta + \sigma_m V_m$$

and

$$E_c = E_f V_f \eta + E_m V_m$$

The approximate values for this factor η are as follows:

$\eta = 1$ for fully aligned and uni-directional fibres
$\eta = \frac{1}{2}$ for bi-directional fibres
$\eta = \frac{1}{3}$ for random in one plane fibres
$\eta = \frac{1}{6}$ for three-dimensional randomised fibres.

Critical Fibre Length

If the fibre is too short it will not transfer the stress from the matrix to the fibre by shear tractions and thus the term ***critical fibre length*** has been introduced. Below the critical fibre length the matrix cannot get a grip on the fibre and under tensile stress the fibre can be pulled out of the matrix. The critical fibre length l_c is expressed as follows:

$$l_c = \frac{\sigma_f d}{2\tau}$$

where σ_f = tensile strength of the fibre
d = diameter of the fibre
τ = interfacial shear strength.

Care should be taken that the fibres are not broken down to less than their critical length during mixing with the matrix.

Mechanical Properties of Typical Composites

The materials used for composites depend upon the service temperature. Resin matrices cannot be used above 300°C owing to degradation of the matrix. Above 300°C metal matrices have to be used. Future developments may well include bi-metallic composites in

Composite material 0·6 V_f	*Relative density* kg/m^3	*Modulus* $N/mm^2 \times 10^3$	*Tensile strength* N/mm^2
E glass/resin	2 050	41	1 240
S glass/resin	2 080	52	1 790
Boron/resin	1 970	269	1 350
Carbon/resin:			
Type 1. High modulus fibre	1 600	190	1 960
Type 2. High strength fibre	1 500	128	1 350

which both the fibre and the matrix are metallic. Fibre-strengthened plastic materials are noted for their high specific strength and high specific modulus, i.e. strength and modulus divided by the relative density. Typical values are shown in the table on page 199.

The chief disadvantages of glass fibres are that their surfaces are sensitive to moisture and flaws and that the elastic modulus is low. The more recent boron fibre and carbon fibre composites give improved specific modulus and specific strength.

EXPANDED PLASTICS AND FOAMS

Many thermoplastics, thermosets and rubbers may be expanded by the formation of gas pockets within the plastic. This produces a cellular structure which may provide one or a combination of the following characteristic properties:

(i) Light weight
(ii) Thermal and electrical insulation
(iii) Energy absorption
(iv) Flexibility

The cellular structure may be of two types, namely:
(i) open cell, i.e. intercommunicating cells or
(ii) closed cell, i.e. non-intercommunicating cells.

Generally the terms *sponge* or *foam* may be used for the open cells and the term *expanded* for closed cells. It is apparent that the open cell structure will expose a greater surface area to attack by gases or fluids.

The most widely used method for making plastic foams is by gas expansion. This can be achieved in several ways, e.g.

(i) Introducing a gas, e.g. carbon dioxide or nitrogen, under pressure directly into the soft plastic mass. The subsequent release of the gas during curing results in a cellular structure.
(ii) Air mechanically mixed with the resin in the uncured state, e.g. urea formaldehyde foams.
(iii) Introducing a low boiling point solvent under pressure. This volatilises when the pressure is released at a temperature above the boiling point of the solvent.
(iv) Introducing a chemical which decomposes, thereby releasing a gas.
(v) Incorporating chemicals in the formulation which react during polymerisation, producing a gas.

Expanded polystyrene can be manufactured by heating impregnated polystyrene beads by steam, causing them to expand into spheres. After a time they are placed in a mould and again heated by steam which fuses the spheres together and increases the degree of expansion. Expanded polystyrene can be used for cushion packs for delicate instruments and for insulation purposes.

Foams have important applications in providing thermal and sound insulation. They may also be used to form light rigid structures by foaming between thin skins. They find applications in the building and aircraft industries. Polyurethane foam is used for making cushions for upholstered furniture and in car seats.

THE ADVANTAGES AND LIMITATIONS OF POLYMER MATERIALS

Advantages

1. Low Density

The densities range from about 800 kg/m³ for polymethylpentene and 900 kg/m³ for low density polyethylene to 2 200 kg/m³ for P.T.F.E. This lightness is utilised in the manufacture of domestic consumer goods. P.T.F.E. also possesses a low coefficient of friction and is therefore a constituent of materials used for dry bearings.

2. Good Electrical and Thermal Insulation

Since the atoms in polymers are joined by covalent bonds there are no free electrons such as exist in metals. Polymer materials are therefore good insulating materials. Controlled conductivity can be obtained by introducing conductive fillers such as carbon.

3. Resistance to Chemical Attack

Polymer materials are generally resistant to chemical attack by a wide range of chemical reagents. The most inert polymers are the fluorocarbons such as P.T.F.E. and fluorocarbon rubber 'Viton'. However, most other plastics are attacked by at least one type of reagent such as strong oxidising acids or organic solvents. Nylon is resistant to most common solvents and to alkalies but is attacked by strong acids. ABS copolymer is soluble in ketones, esters and some chlorinated hydrocarbons. Natural rubber is attacked by minute quantities of ozone in the atmosphere and by many organic solvents. Degradation by oxidation in polymers can be controlled by the addition of anti-oxidants during manufacture.

Polymers avoid the formation of electrochemical cells which are often found in metallic structures. The use of polymers to isolate such electrochemical cells is becoming increasingly important.

4. Surface Finish and Appearance

The surface finish is permanent and requires little or no maintainence. A variety of colours can be obtained by the addition of suitable pigments and dyes. These colours can be moulded in depth and consequently appearance is not affected by surface damage.

5. Ease of Production

Thermoplastics can be readily moulded by the techniques of injection moulding, extrusion and blow moulding. Compression moulding is used for the production of thermoset and rubber articles.

DISADVANTAGES

1. Low Modulus of Elasticity

The modulus of elasticity is lower than that of metals and as previously mentioned varies with strain rate and temperature. Amorphous glassy polymers may have a modulus in the range 10^3 to 10^4 N/mm^2 whilst crystalline polymers are lower by a factor of 10. Typical rubbers have a modulus of about 1 N/mm^2 or less.

2. Liability to Creep at Room Temperature

Creep may occur under prolonged loading at room temperature in polymer materials for highly stressed parts.

3. High Thermal Expansion

The coefficients of linear expansion of polymers are considerably higher than those of metals. This results in difficulties in the design of metal moulds for the moulding of polymers. Differential contraction of the mould and polymer may result in warping in the moulded polymer.

4. Low Maximum Service Temperatures

Most polymers can only be used at relatively low temperatures. Many cannot be used above the boiling point of water. P.T.F.E. and polyimide may be used, subject to stressing conditions, up to approximately 320°C and 400°C respectively.

5. High Notch-sensitivity

Many plastics are notch-sensitive. Care must therefore be taken in design to avoid sudden changes in section and sharp corners.

6. Water Absorption

Some polymers absorb moisture in service with resultant dimensional changes and modifications of their properties, e.g. reduction in the mechanical and insulating properties. Nylon and cellulose acetate have relatively high water absorption rates whilst P.T.F.E. is unaffected.

7. Environmental Stress-cracking

Plastics such as polyethylene are prone to environmental stress-cracking. This is a type of failure by surface-initiated brittle fracture under multi-axial stress when in contact with specific media. Soaps and detergents may cause such failure with polyethylenes of low and medium molecular weights. High density polyethylene has an improved resistance to environmental stress cracking and polypropylene, the latest polyolefine thermoplastic, is immune.

Examination Questions

Questions Nos. 1–65 inclusive are suitable for H.N.C. and H.N.D. students. They are taken from past papers set at the North Gloucestershire College of Technology, Cheltenham and are included by kind permission of the Joint Committee for National Certificates and Diplomas in Engineering.

Questions Nos. 66–83 inclusive have been included as suitable exercises for students of the Mechanical Engineering Technicians course.

CHAPTERS 1–3

1. Describe, with the aid of diagrams, the mechanism of crystallisation of a pure metal. Make neat sketches of the three main types of metallic space lattice, giving examples of metals which crystallise in each type.
2. Explain the fundamental differences between hot-working and cold-working. What are the effects of hot-working on the structure and properties of a mild-steel ingot? Explain the importance of finishing temperature in hot-working.
3. Give an account of the changes which occur during the annealing of cold-worked metals, indicating the factors which affect them.
4. Sketch and explain the microstructure that you would expect upon examination of a specimen of 70:30 cupro-nickel in the cast condition. How would this structure be affected by (1) annealing (2) cold-working after annealing (3) annealing after cold-work?
5. The determination of cooling curves of various alloys of bismuth and antimony gave the following results:

% *Antimony*	*0*	*20*	*40*	*60*	*80*	*100*
1st Arrest°C.	271	400	490	550	600	631
2nd Arrest°C.	—	285	320	370	450	—

Using the above data draw and label the bismuth-antimony thermal equilibrium diagram.

Describe with reference to a cooling curve the cooling of an alloy containing 50% of each metal, estimating:

(1) the composition of the liquid and solid phases at 420°C.

(2) the relative weights of solid and liquid at 420°C.

What is the effect of rapid cooling on the microstructure of the alloy?

6. Bismuth melts at 271°C. and tin at 232°C. They form a eutectic containing 44% tin which melts at 132°C. Bismuth dissolves a maximum of 4% tin, and tin a maximum of 12% bismuth at the eutectic temperature. Draw and label the bismuth-tin thermal equilibrium diagram. Describe with reference to cooling curves the cooling of alloys containing:

(*a*) 60% tin

(*b*) 90% tin from the liquid state to room temperatures.

7. The determination of cooling curves of various alloys of zinc and cadmium gave the following results:

% Cadmium	*0*	*20*	*40*	*60*	*83*	*90*	*100*
1st Arrest°C.	419	382	345	310	266	280	321
2nd Arrest°C.	—	266	266	266	—	266	—

Using the above data, draw and label the thermal equilibrium diagram for this series of alloys.

With reference to a cooling curve, describe the cooling of an alloy containing 30% cadmium, estimating:

(1) The composition of the constituents present at 320°C.

(2) The relative weight of solid to liquid at 320°C.

(3) The proportion of eutectic in the final structure.

8. Two metals A and B have melting points of 960°C. and 800°C. respectively. The metals are completely soluble in one another in the liquid state, but only partially soluble in the solid state. The two branches of the liquidus intersect at 500°C. at a point corresponding to 60%B. At this temperature, the two solid solutions contain 20%B and 85%B respectively, and at 0°C. they contain 5%B and 95%B. Draw and label the thermal equilibrium diagram for this series of alloys.

Describe the changes taking place when:

(*a*) an alloy containing 10%B is slowly cooled from 1,000°C. to room temperature.

(*b*) the same alloy is slowly cooled from 500°C. to 300°C. and then quenched in water.

(*c*) an alloy containing 70%B is slowly cooled from 700°C. to room temperature.

9. Distinguish clearly between the following terms, using diagrams where necessary:
   (*a*) face-centred cubic and body-centred cubic lattice;
   (*b*) eutectic, eutectoid;
   (*c*) substitutional and interstitial solid solution;
   (*d*) liquidus and solidus.

CHAPTERS 4–6

10. Show by means of a graph the effect of carbon content on the mechanical properties of slowly cooled plain carbon steels.

    Compare the microstructures, properties and uses of plain carbon steels containing 0·2%, 0·6% and 1·2% carbon.

11. Give an account of the sulphide inclusions in plain carbon steels.

    Describe how you would determine the distribution of sulphide inclusions along a section of a rolled-steel bar, giving the relevant theory.

12. Discuss the importance of the following factors in the heat-treatment of steel:
    (*a*) furnace atmosphere;
    (*b*) control of temperature;
    (*c*) quenching media and technique;
    (*d*) design of component.

13. Four specimens of 0·6% carbon steel in the form of half-inch round bar are heat-treated as follows:
    (1) Furnace cooled from 800°C.
    (2) Air cooled from 800°C.
    (3) Water quenched from 800°C.
    (4) Water quenched from 800°C., followed by tempering at 600°C.

    Describe the effects produced on the (1) microstructure (2) properties of the steel.

14. Explain briefly, with the aid of diagrams, each of the following terms relating to the heat-treatment of steel:
    (*a*) Overheating
    (*b*) Martempering
    (*c*) Normalising
    (*d*) Underannealing
    (*e*) Sorbite.

15. Sketch and label a typical time-temperature transformation curve (S curve) for a plain carbon steel.

    With reference to this diagram describe the processes of:
    (*a*) martempering and
    (*b*) austempering, indicating the advantages and limitations of these processes.

16. Describe with the aid of diagrams the structural changes that occur during:
    (1) the process-annealing of cold-worked 0·15% carbon steel sheet;
    (2) the full-annealing of a 0·5% carbon steel.

17. Write a short account of the process of surface hardening by pack carburising.

    Outline, with reasons, the heat-treatment necessary to develop the optimum properties in a case-hardened component, indicating clearly the structural changes which occur.

18. Give an account of the process of surface-hardening by nitriding. What are the advantages of this method over pack carburising?

19. Explain clearly the following terms, illustrating your answer with diagrams where necessary:
    (*a*) burnt steel
    (*b*) reducing atmosphere
    (*c*) secondary hardening
    (*d*) critical cooling rate
    (*e*) muffle furnace.

20. Sketch that part of the iron-carbon diagram which is of use in the heat-treatment of steels. With reference to the diagram define the terms: full annealing, normalising, hardening and tempering.

21. What is meant by the term 'hardenability'? What factors influence hardenability and how can it be determined?

    What difference in structure and properties would you expect to find in a 0·4% carbon steel in the form of (1) 12·5 mm diameter bar; (2) 75 mm diameter bar after quenching in water and tempering at 600°C?

22. Describe with the aid of a neat diagram the Jominy end-quench hardenability test. Sketch and explain the microstructures that you would expect to observe along the length of the end-quenched specimen.

CHAPTERS 7–9

23. State the limitations in the use of plain carbon steels. What are the general effects of alloying elements in steel?

24. Compare and contrast the effect of nickel and chromium as alloying elements in steel.
25. What are the general advantages of alloy steels over plain carbon steels? Give an account of the effect of manganese as an alloying element in steel.
26. Give an account of the composition, properties and uses of the low-alloy nickel-chrome steels. What do you understand by the term 'temper-brittleness'? Indicate briefly how this defect may be overcome or minimised.
27. State typical compositions of alloy steels suitable for each of the following:
    (*a*) nitriding
    (*b*) case-hardening
    (*c*) railway points
    (*d*) cutlery
    (*e*) high-speed cutting tools.

    In each case give reasons for your choice and state any heat-treatment which may be required.
28. The nominal compositions of five alloy steels is given as follows:
    (*a*) C=0·1%, Cr=18%, Ni=8%, Ti=0·6%
    (*b*) C=0·3%, Cr=13%
    (*c*) C=0·3%, Cr=1%, Ni=3%, Mo=0·3%
    (*d*) C=1·2%, Mn=12·5%
    (*e*) C=0·6%, Cr=4%, W=18%, V=1%

    State the chief properties and uses of each of the above alloys and describe the heat-treatment necessary to develop their optimum properties.
29. Give an account of the composition, properties and uses of the austenitic stainless steels. What do you understand by the term 'weld-decay'? Indicate briefly how this defect may be overcome or minimised.
30. The strength and shock resistance of grey-iron castings is generally inferior to that of steel forgings. Explain briefly the reasons for this and describe two methods of improving the properties of grey cast iron.
31. Give an account of the production, microstructure, properties and uses of each of the following:
    (1) Grey phosphoric cast iron
    (2) Inoculated high-duty cast iron
    (3) White-heart malleable cast iron.
32. (i) Give an account of the effect of (*a*) rate of cooling and (*b*) chemical composition on the microstructure and properties of ordinary cast iron.
    (ii) Describe the effect of nickel as an alloying addition in cast iron

33. State a typical composition for high-speed tool steel. Describe briefly how you would harden such a steel. Make a neatly labelled diagram of the furnace used for this purpose.

CHAPTERS 10–12

34. Sketch and label the copper-rich portion of the copper-zinc thermal equilibrium diagram.

    Show by means of a graph the effect of constitution on the mechanical properties of the brasses.

    What composition would you recommend for (*a*) deep drawing; (*b*) hot stamping?

35. Make neat sketches of the microstructure of each of the following, indicating clearly the constituents present:

    (*a*) 70:30 brass. Cold-worked + annealed condition
    (*b*) 60:40 brass. Extruded condition
    (*c*) Admiralty gun-metal. Sand cast condition
    (*d*) 80:20 cupro-nickel. Cast condition.

    Indicate briefly the chief properties of each of the above alloys.

36. Describe, with reference to the appropriate portion of the copper-aluminium equilibrium diagram, the heat-treatment of a 10% aluminium-bronze alloy, indicating clearly the structural changes that occur.

    What do you understand by the term 'self-annealing'? Indicate briefly how this defect may be overcome or minimised.

37. What are the characteristic properties of aluminium? Indicate how these affect the uses of the metal and its alloys. Classify the alloys of aluminium, stating the approximate composition, properties and uses of *one* alloy in each group.

38. Give an account of the phenomenon of 'age-hardening' with reference to aluminium alloys of the Duralumin type. State the compositions of two other non-ferrous alloys which exhibit age-hardening.

39. List the characteristic properties and typical uses of the aluminium-bronze alloys.

    What is the effect of the following treatments on the structure and properties of a 10% aluminium-bronze alloy:

    (*a*) Slow cooling from 900°C.
    (*b*) Water quenching from 900°C.
    (*c*) Treatment (*b*) followed by tempering at 600°C?

40. What are the characteristic properties of aluminium-silicon alloys? What do you understand by the term 'modification treatment'? What is the effect of this treatment on the structure and mechanical properties of a 13% silicon alloy?

41. What are the characteristic properties of magnesium? Classify the alloys of magnesium and indicate their more important uses.
42. State the composition, properties and one important use of each of the following:
    (*a*) Monel
    (*b*) K-Monel
    (*c*) Inconel
    (*d*) Nimonic 80.

    Describe briefly the heat-treatment given to (*b*) and (*d*) to develop their optimum properties.
43. Write a brief account of *two* of the following:
    (*a*) Zinc-base die castings
    (*b*) The Nimonic series of alloys
    (*c*) The phosphor-bronze alloys.
44. What are the requirements of an alloy to be used as a bearing material?

    State the composition and properties of a typical (1) copper-base (2) tin-base (3) lead-base bearing alloy. Sketch the micro-structure of any one of the alloys mentioned.

CHAPTERS 13–16

45. Give a brief account of the electro-chemical theory of corrosion. What do you understand by the 'differential aeration effect' in corrosion? Indicate briefly how this explains the phenomenon of pitting.
46. Explain briefly each of the following terms relating to corrosion:
    (*a*) Sacrificial protection
    (*b*) Anodising
    (*c*) Season cracking
    (*d*) Pitting
    (*e*) Sherardising.
47. Give an account of the various methods used to protect iron and steel from corrosion.
48. Discuss briefly the effect of the following factors in the corrosion of metals and alloys, giving examples where necessary:
    (*a*) type of microstructure
    (*b*) natural and artificial oxide films
    (*c*) the presence of internal stress
    (*d*) contact between dissimilar metals.
49. Explain clearly why:
    (*a*) tough-pitch copper cannot be welded satisfactorily;
    (*b*) stabilised stainless steel should be specified for welding purposes;

(*c*) low hydrogen electrodes are used for the welding of low-alloy high-tensile steels;
(*d*) pre-heating is essential for the successful welding of a cracked cast-iron component.

50. What are the essential requirements of an alloy to use as a soft solder? Give an account of the composition, properties and uses of (i) tinman's solder and (ii) plumber's solder.
   What is meant by 'silver soldering'?
51. What are the causes of hard-zone cracking in the welding of low-alloy high-tensile steels? Indicate how this defect may be overcome or minimised.
52. Discuss the effect of hydrogen in the welding of:
   (*a*) aluminium
   (*b*) copper
   (*c*) low-alloy high-tensile steels.
   Indicate briefly how these effects may be overcome or minimized.
53. Discuss the advantages and limitations of the various methods used in the non-destructive testing of metals and alloys.
54. (i) Make neat sketches of the various microstructures that you would expect to observe across a section of a multi-run manual arc-welded joint in half-inch thick mild steel plate.
   (ii) Give reasons for the type of structures observed.
55. Comment on the term 'hardness'.
   State how you would harden each of the following without change in chemical composition:
   (*a*) commercial copper
   (*b*) an aluminium alloy containing 4% copper
   (*c*) a medium carbon steel.
   Describe briefly the principle underlying any one of the above methods.
56. Many service failures have occurred due to brittle fracture in metals which are normally considered ductile. Explain what is meant by the terms 'brittle fracture' and 'transition temperature'. Discuss the factors that effect the transition temperature of mild steel.
57. (i) Describe with the aid of a diagram, the Wohler-type of fatigue test.
   (ii) Sketch and describe a typical S–N curve for (*a*) mild steel and (*b*) a non-ferrous alloy.
   (iii) Discuss the effect of the following factors on the fatigue limit of a steel:
      (*a*) surface hardness
      (*b*) non-metallic inclusions
      (*c*) corrosive environment.

58. What is meant by the term 'creep' in metals and alloys. Sketch and explain a typical creep curve. Discuss the metallurgical factors affecting the creep resistance of a metal.

MISCELLANEOUS

59. (*a*) Describe the chief types of bonding between the atoms in engineering materials.
    (*b*) Indicate how the type of bonding affects the characteristic properties of the material.
60. (*a*) Distinguish between the structure and properties of thermoplastic and thermosetting plastics, giving examples of each.
    (*b*) Describe, with examples, how the properties of resins are affected by the addition of
    (i) plasticisers
    and (ii) fillers.
61. (*a*) Explain clearly the meaning of the term 'glass transition temperature' as applied to thermoplastics.
    (*b*) Describe and illustrate the principal mechanical models of visco-elastic response in polymer materials.
62. Give an account of the following, using diagrams where necessary:
    (*a*) cumulative damage in fatigue
    (*b*) the characteristics of fatigue fractures
    (*c*) the effect of surface treatments on the fatigue limit of a steel.
63. Give an account of the manufacture, properties and chief applications of expanded plastics and foams.
64. Write a short account of *two* of the following:
    (*a*) The effect of nickel in case-hardening steels
    (*b*) Spheroidal graphite cast iron
    (*c*) The isothermal transformation of austenite.
65. Distinguish clearly between each of the following:
    (*a*) Soft solders and silver solders
    (*b*) Fusion and pressure welding processes
    (*c*) Grey and white cast irons
    (*d*) Acid and basic steels.

EXAMINATION QUESTIONS FOR MECHANICAL ENGINEERING TECHNICIANS

The following questions have been selected from specimen and past examination papers for Part II of the Mechanical Engineering Technicians Course by kind permission of the City and Guilds of London Institute. Questions Nos. 66–76 are taken from examinations in the subject 'Engineering Construction and Materials'. Questions Nos. 77–83 are taken from examinations in the subject 'Mathematics and Engineering Science'.

66. (*a*) Explain briefly the difference between malleable and spheroidal graphite cast irons.
    (*b*) Choose an application for which either material might be suitable and outline the factors which would be considered in making a decision which to use. (May 1965).
67. (*a*) Outline the main advantages of sintered materials for use in cutting tools.
    (*b*) Name *two* sintered materials commonly used for cutting tools and indicate, with reasons, the type of application for which they are suitable. (May 1965).
68. (*a*) Give a typical application of a magnesium-base alloy and outline the main properties of this material.
    (*b*) Explain *one* peculiarity of magnesium-base alloys with respect to:

    (i) machining, and
    either
    (ii) directionality
    or
    (iii) electro-chemical behaviour. (May 1965)

69. Aluminium alloys may be grouped roughly according to the main alloying elements, e.g. those using silicon.
    Name any two groups by giving the main alloying element. Explain the properties and state a typical application for each group. (December 1964)
70. (*a*) Give brief details of the general properties of nickel-base alloys.
    (*b*) State *one* application of a particular nickel-base alloy and the properties which fit the application. (December 1964)
71. Give an example of a material in the group known as 'high-tensile bronzes', quoting its approximate composition, main properties, and a typical application. (June 1964)

72. (*a*) By reference to approximate composition and applications, show that stainless steels may be roughly divided into three recognised groups.
    (*b*) For what conditions would a nickel-base alloy such as Monel be preferable to stainless steel? (Specimen Paper)
73. Compare aluminium alloy with cast iron as a material for the manufacture of machine-tool parts. Give an example to show where light alloy has successfully replaced cast iron in this field. (Specimen Paper)
74. (*a*) What is meant by spheroidal graphite cast iron?
    (*b*) Compare briefly, the general properties of spheroidal graphite cast iron and of ordinary cast iron, and show how the uses to which cast iron may be put have been greatly extended by this new form of the material. (May 1963)
75. Give the approximate composition of any common zinc-base die-casting alloy. Explain the advantages and limitations of this material, and give three typical applications. (May 1963)
76. Give a brief explanation to show that the range of high chromium corrosion-resistant steels can be divided into three main groups. Refer to the distinctive properties of each group, and give an example of a typical application of each. (December 1963)
77. Two pieces of mild steel, 19 mm diameter by 150 mm long, were butt-welded to provide an assembly. The assembly failed in service and overheating in the welding process was suspected.

    Describe, in detail, a metallurgical method by which overheating could be verified. (Specimen Paper)
78. Describe how the Brinell and Diamond Pyramid Hardness Tests are performed, and how the hardness numbers are determined in each case.

    Give three reasons why the Diamond Pyramid method has tended to supersede the Brinell method. (Specimen Paper)
79. Describe how

    either (*a*) the Jominy Test

    or (*b*) the Izod Test

    would be carried out on a metal and indicate the information that the test provides. (May 1963)
80. A hot worked steel lacked ductility and it was thought that this was due to the final rolling operations having been conducted at temperatures too low. Describe how to prepare a specimen for metallurgical examination, and the evidence that should prove the contention. (December 1963)
81. Explain what is meant by any *four* of the following metallurgical terms, indicating their connection with the heat-treatment of steel:

(*a*) critical cooling rate,
(*b*) recrystallisation temperature,
(*c*) solid solution,
(*d*) mass effect,
(*e*) limiting ruling section. (June 1964)

82. (*a*) What are the main purposes of introducing alloying elements into steels?
    (*b*) Name a main alloying element of a high-speed steel. For what purpose is the element added, and how does it achieve this purpose?
    (*c*) Why can a high-speed steel be hardened by cooling from an appropriate temperature by means of an air blast?
    (*d*) Why can some hardened high-speed steels have their hardness increased when they are subjected to sub-zero temperatures? (December 1964)

83. (*a*) Distinguish carefully between 'hardness' and 'hardenability'.
    (*b*) Name and briefly describe a test which gives some indication of hardenability.
    (*c*) Sketch the results which the test described in (*b*) would produce with
        (i) a plain 0·9% carbon steel,
        (ii) a nickel-chrome-molybdenum low alloy-steel. (May 1965)

Questions Nos. 2, 3, 10, 11, 12, 19, 20, 26, 30, 33 and 38 are also suitable exercises for Mechanical Engineering Technicians.

Bibliography

1. Institution of Metallurgists, *The Joining of Metals* (lectures delivered at Refresher Course 1951).
2. Institution of Metallurgists, *The Behaviour of Metals at Elevated Temperatures* (lectures delivered at Refresher Course 1956).
3. Institution of Metallurgists, *Year Book and List of Members* 1958–9.
4. American Society of Metals, *Metals Handbook.*
5. Journal of the Institute of Metals, *The History of Magnesium* (paper by MAJOR C. J. P. BALL, July 1956).
6. McGraw-Hill, *Engineering Metallurgy*, MONDOLFO and ZMESKAL.
7. McGraw-Hill, *Applied Metallurgy for Engineers*, BURTON.
8. McGraw-Hill, *Principles of Metallography*, WILLIAMS and HOMERBERG.
9. Arnold, *Metallurgy for Engineers*, E. C. ROLLASON.
10. Arnold, *Introduction to Metallic Corrosion*, U. R. EVANS.
11. Pitman, *Engineering Metallurgy*, SISCO (editor). Committee on Metallurgy (New York).
12. Pitman, *High Temperature Alloys*, CLARK.
13. Pitman, *Microscopical Techniques in Metallurgy*, THOMPSON.
14. Pitman, *Engineer's Approach to Corrosion*, TRIGG.
15. Macmillan, *Text Book of Metallurgy*, A. R. BAILEY.
16. John Wiley; Chapman and Hall (London), *Elements of Heat-Treatment*, ENOS and FONTAINE.
17. Chapman and Hall, *Steels for the User*, ROLFE.
18. Chapman and Hall, *Welding of Non-Ferrous Metals*, DR. E. G. WEST.
19. Institute of Metals, *Non-Destructive Testing of Metals*, DR. R. F. HANSTOCK.
20. British Welding Research Association, *Memorandum of the Non-Destructive Methods for Examination of Welds.*
21. Morgan Brothers, *Kempe's Engineer's Year Book.*

22. Blackie, *Metallurgy*, GREGORY.
23. Pitman, *The Heat-Treatment of Steel*, GREGORY and SIMONS.
24. Butterworth, *Metals Reference Book*, Editor, SMITHELLS.
25. English Universities Press, *Engineering Metallurgy*, Vols 1 and 2, R. A. HIGGINS.
26. Publications by: International Nickel Ltd.; Henry Wiggin and Co., Ltd.; The Aluminium Federation; Copper Development Association.
27. The Association of Light Alloy Refiners and Smelters, *Properties and Characteristics of Aluminium Casting Alloys.*
28. Cooke-Troughton and Simms Ltd., *Photomicrography with the Vickers Projection Microscope.*
29. Kelvin and Hughes (Industrial) Ltd., Publication on *Ultrasonic Flaw Detector, Mark 5.*
30. Griffin, *Process and Physical Metallurgy*, J. E. GARSIDE.
31. Murex Welding Processes, *The Welder*, Coronation number, June 1953.
32. *Journal Iron and Steel Institute*, Vol. 163, p. 277, 1949. Paper by Bardgett and Reeve on Fortiweld Steel.
33. British Standards Specifications.
34. Iliffe, *Metal Industry Handbook and Directory.*
35. Arnold, *Applied Chemistry for Engineers*, GYNGELL.
36. Oxford University Press, *Iron and Steel Today*, J. DEARDEN.
37. Blackie, *An Introduction to the Properties of Engineering Materials*, K. J. PASCOE.
38. Institution of Metallurgists, *Toughness and Brittleness in Metals* (lectures delivered at refresher course 1960).
39. Institution of Metallurgists, *The Structure of Metals* (lectures delivered at the refresher course 1958).
40. McGraw-Hill, *Theory of Metal Cutting*, P. H. BLACK.
41. Iliffe, *Principles of Structural Metallurgy*, B. HAROCOPOS.
42. Butterworths, *Practical Physical Metallurgy*, R. RAWLINGS.
43. Iliffe, *Metallurgical Principles for Engineers*, J. G. TWEEDALE.
44. Macmillan, *Newer Engineering Materials*, Editor, R. F. WINTERS.

Index